Video Compression for Flash, Apple Devices and HTML5

By Jan Ozer

Video Compression for Flash, Apple Devices and HTML5

Jan Ozer

Doceo Publishing
412 West Stuart Drive
Galax, VA 24333

www.doceo.com
www.streaminglearningcenter.com

Screen images courtesy of Rene Marie

ISBN 978-0-9762595-0-3

Library of Congress Control Number: 2011927369
Printed in the United States of America

For Mom

Acknowledgements

This book started out as the written manifestation of the H.264-related seminars that I've been holding at various StreamingMedia conferences over the last few years, then expanded to include live event production and producing for HTML5, as well as other codecs like VP6 and WMV. I couldn't have written this book without the backing of the StreamingMedia team, both for the aforementioned seminars and for various writing assignments that helped me become familiar with the products and technologies discussed herein. So to Eric Schumacher-Rasmussen, Stephen Nathans-Kelly, Dan Rayburn, Dick Kaser and Tom Hogan Jr. and Sr., thank you, thank you, thank you.

I also want to express my appreciation to the vendors who have provided hardware, software and bountiful assistance, including Accordent, Adobe, Apple, BlackMagic Design, Digital Rapids, Google, HP, Inlet Technologies, Kulabyte, MainConcept, Matrox, MediaPlatform, Microsoft, NVIDIA, Rhozet, Seawell Networks, Sorenson Media, Telestream, ViewCast, Wowza Media Systems and I'm sure some that I've forgotten.

This book is the first published by my company, Doceo Publishing. I'm sure there are many rookie rough edges when it comes to presentation, but you should have seen it before my copy editor/proofreader Lucy Sutton got a hold of it. Lucy was my editor at *Millimeter* Magazine, and her contribution is immeasurable.

All that said, I'm unhappy with the output quality of some of the figures and tables in the print version of this book. To provide higher quality output, I've created a free PDF file with all figures and tables that you can download at http://bit.ly/JanOzerbook1. I apologize for any inconvenience.

As always, thanks to Pat Tracy for technical and marketing assistance.

Contents At A Glance

Contents

Chapter 1: Introduction to Streaming Production

I know that you're impatient to begin producing streaming video files, but there are several concepts you need to understand before starting. For this reason, in this chapter, I define terms like codecs, bandwidth, frame rate, data rate and resolution, and then delivery options like streaming, progressive download and adaptive streaming.

Then I look at two topics critical to streaming video production; deinterlacing and producing video with the proper aspect ratio. Next, we'll look at choosing your target delivery platforms, where you'll learn about Flash, HTML5 and Silverlight, and get a look at the resources in the book for producing for those platforms. I'll conclude by introducing you to an Online Video Platform, a third-party service that's ideal for many smaller producers.

Compression and Codecs

Let's start with a discussion of compression and codecs, which are the technologies that shrink your videos to manageable sizes and make streaming possible.

Compression

Video is an amazingly bulky medium, and without compression, it would be impossible to share over the Internet, or even on DVDs or Blu-ray optical discs. Compression technologies shrink the video file size so you can distribute it. All streaming video compression technologies are lossy in nature, which means that upon decompression, the video is an approximation of the original video, not an exact replica. The reason that a lossy technology is necessary, of course, is because lossless technologies—which deliver an exact replica of the original file—don't produce the necessary reduction in file size.

The critical trade-off for all lossy technologies is file size for quality, so the more you compress, the more quality you lose. All other options being identical, a file produced at 800 kilobits per second (kbps) will look better than a file produced at 400 kbps.

Codecs

Codecs are compression technologies with two components: an enCOder to compress the file in your studio or office and a DECoder to decode the file when played by the remote viewer. As the nifty capitalization in the previous sentence suggests, the term codec is a contraction of the terms encoder and decoder (or COmpression and DECompression, depending upon whom you ask).

There are still-image codecs (like JPEG), audio codecs (like MP3) and more video codecs than you could shake a stick at. These include H.264, VP6, Windows Media, WebM and Ogg Theora in streaming markets, with MPEG-2 dominating DVD and Blu-ray production. H.264 and MPEG-2 are huge in the network and particularly satellite markets.

It's instructive to distinguish codecs from development and delivery environments, the latter of which I discuss extensively later in this chapter. For example, QuickTime and Video for Windows are two development environments that support a number of codecs, including DV, MPEG-2, AVCHD and DVCPROHD, just to name a few. If you see an AVI file, you know it's a Video for Windows file, ditto for .mov and QuickTime. But, the video in that file could be encoded with any of a number of codecs.

Similarly, Windows Media (WMV), Flash (FLV, F4V) and QuickTime (MOV) are the most widely used distribution environments, and each format can utilize multiple codecs. For example, both Flash and QuickTime can deliver video encoded with the H.264 codec. HTML5 is a another distribution environment that also supports multiple codecs (H.264, Ogg Theora, WebM).

When producing your streaming files, you need to know which delivery environment you'll be using, since you may prepare the file differently for the different environments. For example, H.264 files encoded for Flash have an .f4v extension, but QuickTime encoded H.264 files with a .mov extension work best for QuickTime delivery.

Ensuring Playback

To ensure that your viewers can play your encoded files, you need to make sure they have the appropriate player installed. For example, with the Flash Player installed, a computer can play H.264 video delivered via the Flash distribution environment, but not H.264 video delivered via Microsoft Silverlight—for that, you'd need the Silverlight player.

It's not as complicated as it sounds; most producers know the distribution environment they're targeting and that their potential viewers need the appropriate player. What's important is that you understand how a codec differs from a development or delivery architecture. Next time someone says, "I'm producing in Flash," you can ask, "Oh, which codec?"

Or if someone commits the incredible gaffe of saying, "I'm producing a QuickTime file," you can respond, "Please don't be so imprecise. Do you mean that you're encoding with a QuickTime-based encoder, or producing an MOV file? And, by the way, QuickTime is a developmental and delivery environment, but not a codec, so which codec are you using?"

Streaming Fundamentals

With compression and codecs known entities, let's discuss some key streaming concepts.

Bandwidth

Bandwidth is the viewer's connection speed to the Internet. To a great degree, this connection bandwidth controls your viewer's ability to retrieve and play video smoothly over the Internet. Higher delivery bandwidths, like those enabled with cable and DSL, allow you to stream higher-quality video to your viewer. In contrast, those poor souls who connect via modem or even ISDN won't be able to view most of the video available on the Internet, at least not in real time. More on that in a moment.

Note that in the early days of streaming video, producers encoded video to meet the bandwidth capabilities of their target viewers. That is, back when most viewers connected via modems, you had to produce postage-stamp-sized video compressed to somewhere south of 28.8 kbps or the viewers couldn't watch it. Today, with most viewers connecting via broadband capable of 1-4 megabits per second (Mbps) or higher, most producers encode their video to meet quality and cost concerns.

For example, if you scan the websites of television networks and/or large corporations, the typical videos average about 640x480 resolution and are produced at 600-800 kbps, even though many viewers have the capacity to watch higher-bit-rate streams. That's because these producers have to pay for their bandwidth and have decided that 640x480 video at 600-800 kbps provides a sufficiently high-quality experience to meet their viewers' needs. In short, back in the day, most producers encoded their files to meet the target bandwidth of their lowest-common-denominator viewer. Today, as bandwidth to the home has increased, choosing a data rate is largely a cost/quality trade-off.

Of course, just when you thought it was safe to go back in the water, video delivery via mobile is coming to the fore. It's a confusing area that needs lots of clarification, but sometime in the 2011-2012 time frame, mobile will become mission-critical for all but the most casual streaming producers (if it hasn't already). Since mobile devices have constrained bandwidths, you'll have to encode your video to stream within those target bandwidths.

Data Rate

Data rate (or bit rate) is the amount of data per second in the encoded video file, usually expressed in kbps or megabits per second. If ESPN distributes its video at 800 kbps, this means that each 1-second chunk of audio and video comprises about 800 kb of data.

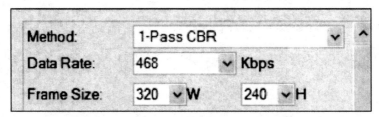

Figure 1-1. Encoding this file to a data rate of 468 kbps.

You set data rate along with other configuration options during the encoding process, but data rate is the most important factor in streaming video quality. That's because, as mentioned, all streaming codecs use lossy compression, so the more you compress, the more quality you lose. For this reason, all other file characteristics (like resolution, frame rate or codec) being equal, the lower the data rate, the lower the quality of the compressed file.

Frame Rate

Most video starts life at 29.97 or 24 frames per second (fps), or 25 fps in Europe. Usually, producers who shoot at 24 fps deliver at that rate, while some producers who shoot at 29.97 fps deliver at 15 fps to reduce the data rate of the encoded file. Though in concept, it feels like dropping the frame rate by 50% would also drop the data rate by 50% with no loss in quality, it seldom works this way. Rather, according to the research that I've performed, the average data rate of video produced at 15 fps is about 20% lower than that of video produced at 30 fps, not 50%. Still a substantial reduction, but often that comes at a subtle quality cost.

For example, when considering 15 fps, note that high-motion video will look noticeably choppy to many viewers. In addition, tight facial shots, where lip synch is critical, often look a bit out of sorts at 15 fps. When deciding which frame rate to use for your video, you should produce test videos at full frame rate and the lower frame rate, and then compare to see which delivers the best overall presentation.

Resolution

Resolution is the height and width of the video in pixels. Most video is originally captured either at 720x480 (standard-definition) or at 1280x720 or 1920x1080 (high-definition), but gets scaled down to smaller resolutions for streaming—usually 640x480 resolution or smaller. This scaling reduces the number of pixels being encoded, making the file easier to compress to lower data rates while retaining good quality.

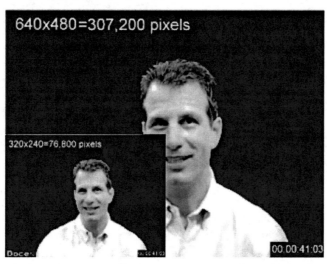

Figure 1-2. Scaling the 640x480 file down to 320x240 reduces the number of pixels by 75%, making the smaller file easier to compress.

For example, a 320x240 video has 76,800 pixels in each frame, while a 640x480 video file has 307,200 pixels, or four times as many, as shown in Figure 1-2. That means you have to apply four times the compression to the 640x480 file to achieve the same data rate as that of the 320x240 file, which usually means noticeably reduced quality. That's why data rate and resolution are integrally linked in quality-related streaming decisions. For example, a video data rate of 250 kbps might look great at 320x240 resolution, but it could look disastrous at 640x480 resolution.

When producing streaming video, you have two options. Option 1 is to choose a data rate, then produce at the highest resolution that looks good at that data rate.

Option 2 is to choose the desired resolution, then output at the data rate necessary to produce acceptable quality at that resolution. The key point is that you should always consider data rate when discussing resolution, and vice versa. Note that the most common video resolutions for 4:3 videos are 640x480, 480x360, 440x330, 400x300, 320x240, 240x180 and 160x120. The most common resolutions for widescreen 16:9 videos are 1280x720, 854x480, 640x360, 480x270 and 320x180.

Distribution Alternatives

It's important to recognize that when you deliver video over the Internet, you have multiple options, including streaming, progressive download and adaptive streaming. Note that the mode you choose may have a significant impact on how you produce your files.

Streaming

Experientially, the concept of streaming means that when a viewer clicks the button on a website, the video starts playing immediately, and it continues to play more or less smoothly to the end. To make this happen, the data rate of the encoded file must be smaller than the bandwidth capacity of the remote viewer; otherwise, the video will frequently stop playing.

That's the experiential definition of "streaming." The technical definition of "streaming video" is video delivered via a streaming server, which is a software program that's charged solely with delivering streaming media. This is in contrast with a traditional web server that runs all websites, and delivers all forms of web content, including HTML text, JPEG and GIF images, PDF files, and the like.

If you're producing for a streaming server, understand that some servers have specific requirements for the streaming files that they deliver; for example, files delivered via the Apple streaming server must be "hinted" for streaming, which is a option available on most encoding programs that produce QuickTime video. In addition, some producers prefer to encode streaming video via constant bit rate (CBR) encoding, which I'll define in the next chapter.

So when producing files for delivery via streaming, you should:

- Encode at a data rate that's comfortably below the connection bandwidth of most target viewers

- Identify any specific requirements for files distributed via the streaming server

- Consider encoding using CBR data rate control.

Progressive Download

The technical definition of "progressive download" is video delivered by a regular HTTP web server rather than a streaming server. In most instances, video delivered using this technique is stored on the viewer's hard drive as it's received, and then it's played from the hard drive. In contrast, streaming video is usually not stored (also called cached) locally, so if the viewer can't retrieve and play it in real time, he or she can't play it smoothly at all.

Experientially, most files delivered via progressive download seem like streaming in that they begin to play once you click the button, and they continue smoothly until the end. For example, YouTube, ESPN and CNN all deliver without a streaming server. However, when delivering via progressive download, you can encode at higher rates than you could when delivering via streaming, because even if the data rate exceeds the viewer's connection bandwidth, ultimately, it will reside on the viewer's hard drive, from which it will play smoothly.

Note that one of the key reasons that producers use streaming servers is because once video is stored (or cached) on a hard drive, it's very easy to copy. Streaming video can be cache-less, which makes it inherently more secure.

Again, as with files produced for streaming, note that some progressive download technologies also have specific requirements. For example, files delivered to the QuickTime Player must be encoded with the Fast Start option selected, while H.264 files distributed to the Flash Player must have the MOOV atom located at the start of the bit stream. Otherwise, using both technologies, the entire file must download before it starts playing. I talk more about both of these options in Chapter 5.

Finally, most producers who deliver via progressive download produce their files using variable bit rate (VBR) techniques, which I'll also define next chapter, since it delivers the optimum blend of file size and quality.

To summarize, when producing for delivery via progressive download, you should

- Choose a data rate that strikes a good balance between quality and waiting time
- Adhere to any technology-specific requirements for progressive delivery
- Consider encoding your files using VBR data rate control.

Adaptive Streaming

Adaptive streaming technologies encode multiple live or on-demand streams and switch them adaptively based upon changing line conditions and other variables. When the connection is good, the viewer gets a high-quality, high-data-rate stream, but if connection speed drops, the server will send a lower-data-rate file to ensure a continuous connection, albeit at lower quality. Adaptive streaming provides the best of all possible worlds: great quality-video for those with the connection speed to retrieve it (and the CPU required to play it back), and a passable-quality stream for those with Wi-Fi, mobile or other slow connections on lower-power devices.

There are multiple adaptive streaming alternatives today, including Adobe's Dynamic Streaming, Apple's HTTP Live Streaming and Microsoft's Smooth Streaming. Chapter 7 provides more details regarding the various adaptive streaming technologies, and describes the optimal procedures for encoding for each technology.

Streaming Production

Beyond these compression- and streaming-related topics, there are two production-related issues that you should know about to produce high-quality streaming video in any format. These are deinterlacing and aspect ratio. Let's take them in that order.

Deinterlacing

Have you ever seen a streaming video that looked like Figure 1-3? I call those slices Venetian blind artifacts, and they are typically caused by video shot in interlaced mode with either inadequate or no deinterlacing. Interestingly, this frame is from a trailer for the movie *Little Miss Sunshine*, which was shot in either film or progressive video, so the interlacing probably appeared after the video was converted to interlaced for DVD or VHS production.

Figure 1-3. This video frame clearly wasn't deinterlaced.

You can see a subtler deinterlacing artifact in Figure 1-4, which is from a Wal-Mart online video advertisement. Specifically, the jaggy lines on the table and back of the computer indicate that the video was shot in interlaced mode and a poor-quality deinterlacing filter was used. Note how the rest of the video is sharp and clear—it's just the diagonal lines that are affected.

Where do these problems come from? Here's the *CliffsNotes* version:

- Most video was originally shot in interlaced format. This means that each frame consists of two fields, each shot 1/60 of a second apart (that's in NTSC-land; in Europe it's 1/50 of a second apart).

- All streaming files are progressive. This means that the two fields must be combined into a single frame.

Figure 1-4. This video frame was either inadequately deinterlaced, or wasn't deinterlaced at all.

• When combining the two fields, the most classic problem occurs in high-motion shots when there's a noticeable difference in the position of the subject in the two fields. That's what you see in Figure 1-5. During the swing, the club head moved about a foot in the 1/60 of a second between when the first and second fields were shot—hence the double image when you combine the fields without deinterlacing.

Figure 1-5. Two fields combined without deinterlacing.

Deinterlacing filters combine the two fields and apply algorithms to minimize the resulting (usually jaggy) artifacts. You can see this on the left in Figure 1-6. On the right, you can see the same frame, shot with a camera in progressive mode that was mounted alongside the camera shooting in interlaced mode.

These figures lead to several deinterlacing-related conclusions. First, the quality of deinterlaced footage will never be as good as that of progressive footage of the same source. When shooting for streaming, always shoot progressive when it's available.

Second, when working with interlaced source footage, it's always deinterlace your source footage before or during the encoding process. If you don't, you'll see artifacts like those in Figure 1-3 in higher-motion sequences.

Figure 1-6. On the left, two fields with deinterlacing; on the right, a progressive frame that looks cleaner.

Finally, if you see artifacts like those in Figure 1-4, deinterlacing quality isn't as good as it could be. Experiment with different deinterlacing settings offered by your encoding tool until you find the one that works best with your source footage. For example, Figure 1-7 shows the options available in the Sorenson Squeeze encoder, with the default (and best) option selected.

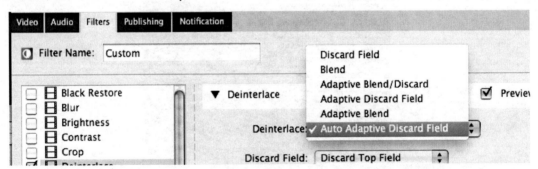

Figure 1-7. Deinterlacing options available in Sorenson Squeeze.

Overall, when working with interlaced source footage, be on the lookout for interlacing artifacts. If you see them, try different deinterlacing settings in your encoder, and don't be afraid to change deinterlacing tools to produce the optimal result.

Aspect Ratio Issues

Aspect ratio issues pop up in the most surprising places. They are evidenced by a mismatch between the appearance of a streaming video file and a digital photo of the same subject. For example, check out CNN's Anderson Cooper in Figure 1-8. On the left is a video frame grab from a CNN streaming video file; on the right, a digital picture from the CNN site. Obviously, the aspect ratio is off on the left.

Figure 1-8. A clear aspect ratio mismatch between the streaming video and a digital picture.

This also happens frequently with corporate streaming videos, as shown in Figure 1-9, which is a frame from an Accenture video, with the inset a digital shot of the same executive.

Figure 1-9. Same issue here in an Accenture video.

The obvious first step in resolving these issues is to recognize that you have a problem. The path toward resolving the problem depends on whether your source footage is standard-definition (SD) or high-definition (HD). Let's examine each in turn.

Resolving SD Aspect Ratio Issues

When working with SD footage—whether 4:3 or 16:9, NTSC or PAL—understand that the video was designed to be viewed on a television set and has a fundamentally different aspect ratio when viewed on a computer. You can see this in Figure 1-10, which shows a DV frame grab displayed on a computer on the left, and the same frame displayed on the television on the right. Essentially, the frame is horizontally squeezed about 10% when displayed on a TV set.

Figure 1-10. SD video displays differently on computers and TV sets.

If you analyze a standard-definition DV file in Adobe Premiere Pro, you'll see that it has an aspect ratio of .9091, which is the circled number in Figure 1-11.

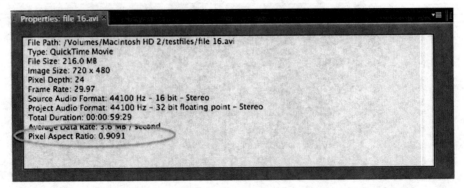

Figure 1-11. A standard-definition 4:3 DV file has an aspect ratio of .9091.

In essence, this means that each horizontal pixel in the video must be squeezed to .9091 of its original size to appear normal. That's about 10%.

Television sets do this automatically, which is why the frame on the right in Figure 1-10 looks skinnier than the frame on the left. However, when producing streaming video, you have to squeeze the video by the same 10% (or, in the case of 16:9 video, expand it by about 20%). Fortunately, you don't have to do this math in your head. Though the procedure will vary by encoding tool, the general rules are the same.

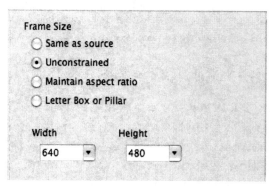

Figure 1-12. Changing the aspect ratio of SD source video in Sorenson Squeeze.

Specifically, here's the procedure:

- Insert the desired frame size in pixels. With 4:3 source footage, the frame size should conform to 4:3, which means 640x480, 480x360, 400x300, 320x240 and so on. With 16:9 source footage, the frame size should conform to 16:9, which means 640x360, 480x270, 320x180 and so on.

- Here's the hard part. With whatever tool you're using, make sure that the aspect ratio is changed from the source to the output. With Squeeze, you select Unconstrained, since Maintain Aspect Ratio would obviously maintain the source aspect ratio, which we don't want, while Letter Box or Pillar will insert black bars into the video, which we also don't want.

- With Apple Compressor, shown in Figure 1-13, in the Geometry Pane, you input the desired frame size and then choose Square (or 1.0000) for the Pixel Aspect ratio. You don't want to choose NTSC CCIR 601/DV for 4:3 footage, or NTSC CCIR 601/DV (16:9) for widescreen footage, because that will maintain the same aspect ratio, and with SD source footage, you want to change it.

For a much deeper look at the issue, check out "Choosing the Optimal Video Resolution" at bit.ly/optimalres. That's it for SD; now let's look at HD.

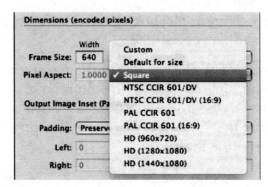

Figure 1-13. Changing the aspect ratio of SD source video in Apple Compressor.

Resolving HD Aspect Ratio Issues

With HD source footage, the solution is much simpler because there is no aspect ratio mismatch—the video should look exactly the same on a computer and television set. This means that you don't have to change the aspect ratio as part of the encoding process. Typically, if you use the Square output shown in Figure 1-13, or the equivalent in your encoding program, you should avoid any aspect ratio mismatches.

Again, fixing aspect ratio issues starts by recognizing that you have a problem. Whenever a video frame doesn't look the same as a digital picture of the same subject, you definitely have a problem. Streaming video shouldn't make a person look fatter or skinnier than a digital photo—he or she should look exactly the same.

Now onto the final major topic I'll cover in this chapter: choosing your delivery targets.

Choosing Your Delivery Environments

Most readers are probably too young to remember Steven R. Covey's *Seven Habits of Highly Effective People*, but it's sold more than 15 million copies in 38 languages, making it one of the best selling self-help books of all time. Habit number 2, which has stuck by me since reading the book back in 1989 (I was 4) is, "Begin with the end in mind." In streaming production, this translates to, "know your delivery environments before you encode your first file." So that's what we're going to discuss.

There are a lot of potential viewers out there, watching on desktop and notebook computers, netbooks, smartphones, tablet computers and other mobile devices, set-top boxes, and other players. Your job is to efficiently reach the most viewers with the highest-possible-quality video. To do that, you have to choose the delivery environments used to access each type of viewer.

For example, for delivering to computers, you have Flash, Silverlight, QuickTime, Windows Media and HTML5. For Apple iOS devices, you have QuickTime and HTML5. For Android, Flash may work, as might HTML5. There are lots of options; each one has certain costs and benefits—you get the point. So in this section, I'm going to lay out my vision of the factors you should consider when choosing your delivery environments. But let's start with a brief history of the streaming market.

A Brief History of Streaming Video

In the beginning, there was RealNetworks, the true pioneer of streaming media. Early on, RealPlayer was the one player you absolutely had to have to enjoy streaming media, even though you absolutely hated its intrusiveness and continual nagging to upgrade to the paid version. This was true through the early '90s.

Then, Windows Media took over. Microsoft started giving away the streaming servers that RealNetworks sold to stay afloat, and perhaps pursued other anti-competitive practices that led to a $761 million antitrust settlement to RealNetworks in 2005. Whatever the reason, Real declined and the bulk of video on the Internet was distributed in Windows Media format.

Flash Rules

In the early 2000s, however, Macromedia Flash was becoming *the* design environment for web pages—first adding simple animations, then complete interfaces, then pretty awful video. Then, in 2005, Macromedia added On2 Technologies VP6 codec, which delivered equivalent quality to Windows Media. High-profile websites that were already using Flash to design their web pages dropped Windows Media like a bad habit, and by 2009 or so, you'd be hard-pressed to find any three-letter network or other prominent business-to-consumer (B2C) site using anything but Flash.

Why Flash? In addition to the fact that it was a great design tool, the Flash Player was ubiquitous on Windows, Mac and Linux platforms. Microsoft did an awful job with Mac support for Windows Media, and Linux was even worse. Unfortunately for Microsoft, most web designers used Macs, suffered under Microsoft's continued lack of love, and switched over to Flash as soon as it was viable.

Silverlight Leaves the Gate Slowly

In 2007, Microsoft launched Silverlight, which, very much like Flash, was both an authoring environment and a video playback environment. If you want to read up on how the two technologies compared in 2008, check out "Flash vs. Silverlight" (bit.ly/flashvsilverlight). The bottom line was that while Silverlight has enjoyed some impressive design wins, including both the Summer and Winter Olympics and Sunday Night Football, the Silverlight player never achieved the penetration necessary for average folks like you and me to use it.

As I write this in March, 2011, Silverlight's penetration is up to 72%, compared with 96% for Flash (see www.riastats.com). That works if you have compelling content like Lindsey Vonn on the downhill, but probably not for your average sales or marketing video.

There are a some scenarios where Silverlight makes a lot of sense—primarily when an organization has significant legacy content in Windows Media format, which won't play in Flash, and doesn't want to re-encode. In addition, when an organization has lots of .NET programmers around who can easily transition over to Silverlight development, that's a better option than retraining them all for Flash development.

That said, both scenarios work best in an intranet or campus scenario where the organization can dictate which plug-ins the most relevant target viewers must have. In contrast, if you're delivering general-purpose content to the general Internet community, you run a significant risk that many viewers simply won't download the plug-in to view your content.

The iPad Rocks Adobe's World

So Flash dominated until early 2010, when Apple introduced the iPad-notably without Flash support. Instead, Apple used HTML5 technology to deliver video to its soon-to-become-very-high-profile device. Then, because the press loves a hot story, we started seeing headlines like "Flash is Dead—HTML5 goodness around the web" and "Is Flash Dead Yet?" Pretty ludicrous given that the penetration of HTML5 browsers at the time was less than 30% or so, which meant that 70% of web viewers couldn't even watch HTML5 video, but, hey, never let facts get in the way of a hot story or a good rant.

Twelve months later, Flash is not dead, but HTML5 is definitely having an impact on how video is distributed over the Internet. So let's take a look at HTML5 and see what the ruckus is about.

About HTML5

HTML stands for "hypertext markup language," and HTML is the primary language used to produce most websites. HTML is a standard set by the World Wide Web Consortium (WC3), a group that includes browser developers and other interested technology companies. HTML5 is the latest iteration of HTML to be introduced by the WC3.

New in HTML5 is a video tag (see Figure 1-14) that enables browsers to natively play back video within the page, obviating the need for plug-ins like Flash, Silverlight and QuickTime. Simpler is always better, and if you don't need a plug-in to play a video file, that's one less item for potential viewers to download.

```
<video class='sublime' height='340' id='single_video' poster='/demo/dartmoor.jpg?1298286295'
<source src='http://medias.jilion.com/sublimevideo/dartmoor.mp4' />
<source src='http://medias.jilion.com/sublimevideo/dartmoor-mobile.mp4' />
<source src='http://medias.jilion.com/sublimevideo/dartmoor.webm' />
<source src='http://medias.jilion.com/sublimevideo/dartmoor.ogv' />
</video>
```

Figure 1-14. HTML5 video tag at work at **sublimevideo.net/demo**. *Why so many files? Read on.*

At a high level, however, you're transferring the burden of innovation from one vendor—Adobe for Flash, Microsoft for Silverlight—to multiple browser vendors: Apple, Google, Microsoft, Mozilla and Opera. For Adobe to add a new feature or codec in Flash, they add the feature and update the player. To update an HTML5-related feature means that you have to gain full support from five different companies on all supported operating systems and browsers.

I've written a bunch on HTML5, most comprehensively in "Five Key Myths about HTML5" (**bit.ly/html5myths**), so I'll be brief. If you're considering HTML5 in the short term (say, prior to 2014), here are some thoughts to consider:

1. *HTML5-compatible browser penetration is still low.* As of early 2012, HTML5-compatible browsers represent about 50% of all browsers. To reach the others, you'll have to fall back to Flash or other plug-in-based technology. It's not difficult to do, but unless HTML5 is delivering some benefit you can't get in Flash, why bother?

2. *Multiple files required.* The WC3 hasn't specified a single HTML5-compatible codec, so you have to produce in at least two—WebM and H.264—to play in all current browsers, and a third—Ogg Theora—to play in older HTML5-compatible browsers. That's why there are three files in Figure 1-14, plus one for mobile delivery. For the record, H.264 plays natively in current versions of Apple Safari, Google Chrome (for the time being; Google announced in early 2011 that it will remove H.264 playback from Chrome in later in the year) and Microsoft Internet Explorer 9. WebM plays natively in current versions of Google Chrome, Mozilla Firefox and the Opera browser.

3. *Features are limited.* HTML5 doesn't offer many features currently enabled by Flash or Silverlight, like adaptive streaming, digital rights management, live streaming, DVR functionality, multicasting (one video stream serves many viewers) and peer-to-peer delivery (launched by Adobe with Flash Media Server 4). HTML5 is OK for simple video playback in a window, but otherwise it's way behind, and it's definitely not moving very fast.

4. *It's going to cost you.* You'll have to rewrite all Flash or Silverlight apps built for your current website into JavaScript to get equivalent functionality for HTML5.

5. *Playback performance may drop.* Because Flash and H.264 are accelerated by many graphics chips on notebooks, computers and mobile devices, HTML5 video—particu-

larly WebM—will require more CPU cycles to decode than Flash and H.264 on many platforms.

As I concluded in the aforementioned "Myths" article:

> HTML5's value proposition today, and for the foreseeable future, is, "Encode in more formats that offer no advantage over H.264, and play on fewer computers, and distribute your on-demand content to vastly fewer viewers with lower quality of service, less features and a reduced ability to monetize than you can with Flash or Silverlight. Oh, and forget live."

Recall my initial statement that, "Your job is to efficiently reach the most viewers with the highest-possible-quality video." In the short term, other than Apple iOS devices—which are really a subset of HTML5, not the whole enchilada—HTML5 doesn't deliver any target viewers that you couldn't otherwise reach via Flash, and the quality of service and feature set is very much inferior. In my view, you have more important priorities in the short term than HTML5.

Not that you can ignore it; HTML5 has to be on your technology agenda. It's just that Flash isn't going anywhere in the short term.

Mobile Rises

In addition to the Apple iOS platforms, mobile is increasing in importance as a relevant target delivery environment. Apple has done the best job supplying technologies like HTTP Live Streaming and documenting how to deliver to its devices, but Android, HP webOS, Nokia, BlackBerry and Windows Mobile are still important targets.

Prioritizing Your Delivery Targets

So what does this add up to? In terms of overall priorities for most general-purpose video producers, your strategy should be:

1. Supporting Flash on computers and all available mobile devices with both single file playback and adaptive streaming. If you're currently supporting Silverlight, I'm sure you have your reasons, and switching to Flash probably isn't likely, so move on to number 2.

2. Supporting iOS devices with single file playback and adaptive streaming.

3. Developing a cohesive strategy for delivering to Android and other mobile devices.

4. Keeping a long-term eye out for HTML5.

With our priorities set, let's see what it takes to produce video for each delivery environment.

Producing for Your Delivery Targets

Let's take a 50,000-foot view of what it takes to produce for each of the aforementioned environments, and then kick you loose to start encoding.

Producing for Flash

The Flash Player supports multiple codecs, including Sorenson Spark, On2 VP6, H.264 and, sometime down the road, WebM. Spark is obsolete for most readers, so I won't discuss that. I detail the various H.264 configuration options in Chapter 4, and detail how to produce H.264 for Flash in Chapter 5. In terms of multiple-file H.264 adaptive streaming, I cover Flash-based RTMP Adaptive Streaming and HTTP Adaptive Streaming in Chapter 7.

If you're still producing in VP6 format, you can learn more about that in Chapter 9. I also cover WebM production in that chapter.

Producing for Windows Media and Silverlight

If you're delivering video to the Windows Media Player, Windows Media Video is your only option, and you can read about that in Chapter 9. If you're producing for Silverlight, you can encode in Windows Media Video or H.264. Again, I cover the various H.264 configuration options in Chapter 4, and I discuss how to produce H.264 for Silverlight distribution in Chapter 5.

If you're interested in producing for Smooth Streaming for Silverlight, that's covered in Chapter 7.

Producing for HTML5

With HTML5, you have three codec options, including WebM (covered in Chapter 9) and Ogg Theora, which I discuss but don't detail in Chapter 9. As above, I detail the various H.264 configuration options in Chapter 4, and I discuss how to produce H.264 for HTML5 in Chapter 5.

Producing for Apple Devices

Chapter 6 is dedicated to Producing for Apple Devices, with a section on Apple's HTTP Live Streaming technology in Chapter 7.

DIY or OVP

Once you decide which environments to target, you have to decide whether you're going to do the work yourself (do it yourself, or DIY) or use a third-party service provider. These include online video platforms (OVP) like Brightcove or Ooyala, or even user-generated content (UGC) sites like YouTube or Vimeo. If you DIY, you (or someone in your organization) have to install, configure and maintain any necessary streaming servers, encode your video, design and create the player, and host and distribute the streaming video files.

As I explain in Chapter 10, if you use an OVP or UGC site, it will do most of this work for you. You upload a high-quality video file to the site and choose some player options, and it encodes the video into the required format, creates the player, and hosts and delivers the video. If you want to start streaming in 20 minutes or less, third-party service providers are a great option that I cover in Chapter 10: Distributing Your Video.

Conclusion

After reading this chapter, you should feel comfortable with most general streaming concepts, and two production issues that can degrade the quality of your streaming video. You should have some insight into which video platforms to target, and the requirements for doing so.

In the next chapter, we cover general encoding concepts that are relevant irrespective of the compression technology that you decide to use.

Chapter 2: Universal Encoding Parameters

With these streaming basics under our belt, let's tackle some universal parameters that relate to producing streaming files in all formats for all target environments.

Constant vs. Variable Bit Rate Encoding

Constant bit rate (CBR) encoding and variable bit rate (VBR) encoding are two techniques for controlling the data rate of a compressed video file.

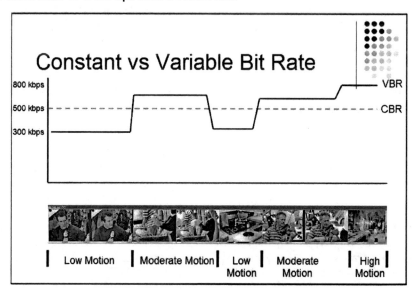

Figure 2-1. Constant and variable bit rate encoding, in theory.

Simply stated, encoding via CBR produces a file that has a constant bit rate throughout. In contrast, encoding via VBR varies the data rate according to the complexity of the video file, while achieving the same average data rate as that of a file produced with CBR.

This is shown in Figure 2-1, which illustrates a file with low-motion, easy-to-compress scenes, and high-motion, hard-to-compress scenes. Both techniques achieve the same average data rate over the duration of the file, but the CBR line, with dashes, stays constant throughout, while the solid VBR line varies with the amount of motion in the scene.

In general, VBR should produce a higher-quality file than CBR because it allocates data rate as necessary to maximize quality. The primary downside is stream variability, since the per-second bit rate can vary significantly from section to section. More on this in a moment.

Producing Optimal-Quality CBR Files

When producing files with CBR encoding, options will vary depending upon whether you're streaming live or creating on-demand files. If you're creating on-demand files, some encoders let you choose between one-pass and two-pass encoding, as shown in Figure 2-2.

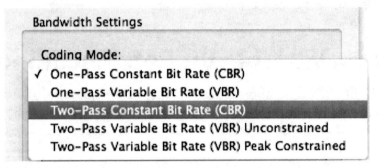

Figure 2-2. Choosing between one-pass and two-pass CBR in Telestream Episode.

What happens with two-pass CBR? Here's a blurb from Microsoft's website in an article titled "Encoding Methods" (**bit.ly/2-passcbr**).

> Standard CBR uses only a single encoding pass. You provide your content as input samples, and the codec compresses the content and returns output samples. It is also possible to process input samples twice. On the first pass, the codec performs calculations to optimize encoding for your content. On the second pass, the codec uses the data gathered during the first pass to encode the content.
>
> Two-pass CBR encoding has many advantages. It often yields significant quality gains over standard CBR encoding without changing any of the buffering requirements. This makes this encoding mode ideal for content that is streamed over a network. The only situation where two-pass CBR is not feasible is when you encode content from a live source and cannot use a second pass.

For these reasons, when two-pass encoding is available in an encoding tool (not all tools enable this for all codecs), I always select it. As the Microsoft quote says, the only time you absolutely shouldn't elect two passes is when you're encoding a live stream.

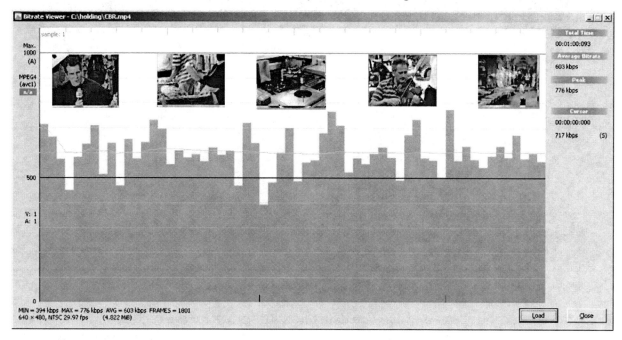

Figure 2-3. A file encoded using CBR encoding. Note how the average data rate line remains very constant.

Figure 2-3 shows a data rate graph of the file shown in Figure 2-1 produced using CBR encoding as displayed in a tool called Bitrate Viewer (which you can read about in Chapter 13). You can see the low-motion sequence at the start, moderate-motion next, and so on (I inserted thumbnails from the video into the image as a reminder). In the figure, the spikes represent the data rate for each second of video, while the faint wavy line represents the overall average data rate. You can see that while there are individual spikes in the graph, the average is pretty constant.

This figure shows that CBR doesn't mean a total flat line; there will be spikes in the data rate. However, when you compare this graph to Figure 2-4, which was encoded using VBR, you'll see much more variability in the VBR file.

Producing Optimal-Quality VBR Files

Figure 2-4 shows the same file encoded using VBR. Unfortunately, the scale is different for the two images, primarily because there were data rate spikes in the VBR file that extended beyond 1 Mbps. But if you ignore this and concentrate on the per-second spikes and faint average data rate line, you can see that both are low for the talking head portion of the file, then boost significantly for the moderate-motion sequence of a Druze woman tossing a pita.

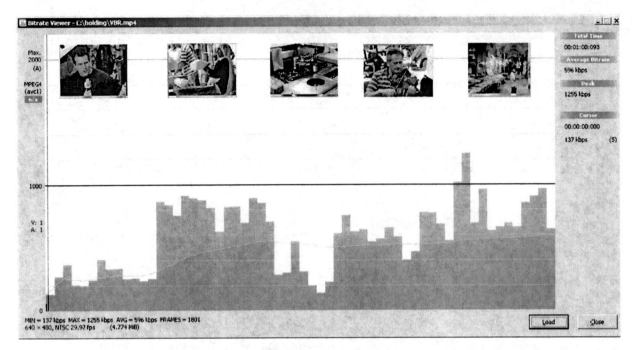

Figure 2-4. A file encoded using VBR encoding.

Then, the per-second spikes and average data rate settle down for the section showing an in-tegrated circuit cutting machine, then increase for the violinist, and peak for the high-motion walk through of a street in Jerusalem. In short, the per-second data rate and average data rate follow the amount of underlying motion in the video file.

So that's VBR, and now we'll discuss concepts relating to VBR production. First is the concept of constrained vs. unconstrained. If you look back at Figure 2-2, you'll see options for Two-Pass Variable Bit Rate (VBR) Unconstrained, and Two-Pass Variable Bit Rate (VBR) Peak Constrained.

The difference is this: With unconstrained VBR, you set a quality level and the encoding tool determines the necessary data rate to meet that quality level. While this is appropriate for archival applications, when you're streaming, you need to control data rate more than quality. So when you're encoding for streaming and see both of these options available, choose Peak Constrained (or simply Constrained), which limits the maximum data rate (and sometimes both the minimum and maximum data rate) to parameters that you set using controls like those shown in Figure 2-5.

Most encoding tools let you set the maximum data rate, though some—like Sorenson Squeeze, shown in Figure 2-5—let you set maximum and minimum. Which values should you use? Figure 2-5 shows my rule of thumb, which is 50% of the target for the minimum data rate and 200% of the target for the maximum.

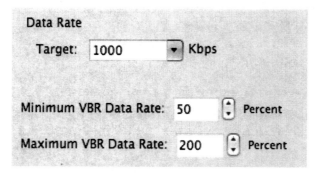

Figure 2-5. Selecting the target data rate, as well as minimum and maximum.

The next VBR-related concept is the number of passes. Some encoding tools limit you to two passes; others enable multi-pass. With two-pass encoding, the encoder scans the video file in the first pass to identify the various sequences in the video file and catalog their complexity, and then encodes the file in the second pass.

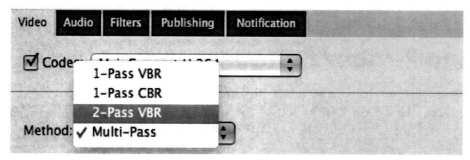

Figure 2-6. Choosing between two-pass VBR and multi-pass.

With multi-pass VBR encoding, the encoding tool uses the additional passes to fine-tune the data rate of the file. For example, in the first pass, the encoder would scan the file, then perform a test encode in the second. If a scene proved harder or easier to compress than originally predicted, the encoder would re-encode using different parameters in the third pass.

The downside of multi-pass encoding is processing time, which can get quite lengthy as some encoding tools use five passes or more. If you're in a hurry, you can run some test encodes to gauge how quality and processing time compare, and see if the extra time is worth it. For example, I've run that analysis with Sorenson Squeeze and learned that the quality difference was negligible, but multi-pass encoding can take about twice as long. As a result, my default for Squeeze at this point is two-pass, not multi-pass.

Choosing Between VBR and CBR

If you shook a compressionist awake in 2007 and asked when to encode with CBR and when to use VBR, the rote answer would be to use VBR when encoding for progressive download, and CBR for streaming. The logic behind this is that when a streaming server is doling out the video, it works more efficiently with a CBR file.

However, when I polled compressionists for a recent article, I learned that most producers used constrained VBR in most settings, irrespective of whether the files were for streaming or progressive download. So my general recommendation is to do the same, with these exceptions:

•   When producing live events with a fixed outbound bandwidth to the streaming server, use CBR. For example, the outbound bandwidth from my office is 850 kbps. When I stream live, I make sure to select CBR. If my outbound bandwidth was 10 Mbps, rather than 850 kbps, I'd consider using constrained VBR.

•   When producing for adaptive streaming, the issue is tougher. See my recommendations for this in Chapter 7.

I-, B- and P-frames

I'll review these materials for H.264 much more extensively in Chapter 4, but I wanted to get these high-level concepts out on the table as early as possible. All codecs use different frame types during encoding. Some, like VP6, use two types—I-frames (also called key frames) and P-frames—while others like H.264 and VC-1 use three types: I-, B- and P-frames.

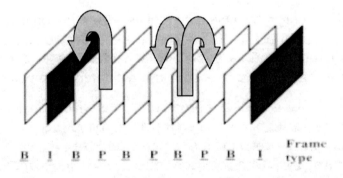

Figure 2-7. The three frame types used during encoding. P-frames look backward for redundancies, while B-frames look forward and backward.

Figure 2-7 shows all three frame types in a group of pictures (GOP), or a sequence of frames that starts with a key frame and includes all frames up to, but not including, the next key frame. Briefly, an I-frame is entirely self-contained and is compressed solely with intra-frame encoding techniques—typically a technology like JPEG, which is used for still images on the web and in many digital cameras.

P- and B-frames are difference frames that refer to other frames for as much content as possible. Imagine a talking head video like that shown in Figure 2-8. When producing a P-frame, the encoder will look back to a previous key (or P-frame) for regions in the frame that haven't changed, like the wall behind the speaker and most of his body and face. Then, it will encode only what's changed between the two frames. During playback, the player displays all pixels from the reference frame except the changed regions.

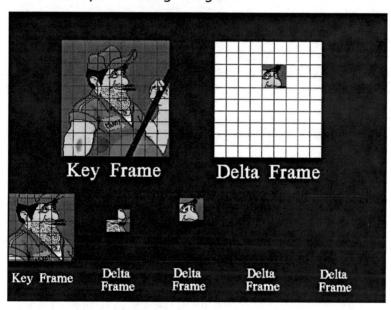

Figure 2-8. Delta frames (B- and P-frames) store the blocks that change from the reference frame (in this case a key frame).

This is why talking head videos compress so efficiently; there's so much inter-frame redundancy that the P- and B-frames contain very little new information, making them easy-to-compress. In a fast-paced soccer game, P- and B-frames contain much more original content, which makes compressing down to the target data rate much tougher.

Back to our frame types. By definition, a P-frame looks backward to a previous P- or I-frame for redundancies, while a B-frame can look backward and forward to previous or subsequent P- or I-frames. This doubles the chance that the B-frame will find redundancies, making it the most efficient frame in the GOP.

Working With I-frames

How do you use these frame types to your advantage? With I-frames, recognize that these are the largest frames, which makes them the least efficient from a compression standpoint. Basically, you only want I-frames where they enhance either quality or interactivity.

For example, video playback must start on an I-frame since B- and P-frames don't contain sufficient content to reproduce the frame. If a viewer drags the slider on her video player to a B-frame, for example, the player must scroll back to the nearest I-frame, and then start decoding until it arrives at the B-frame. To make the video file responsive to viewers navigating via the slider, or otherwise jumping around the video file, I recommend adding a key frame every 10 seconds, or every 300 frames in a 29.97 fps file.

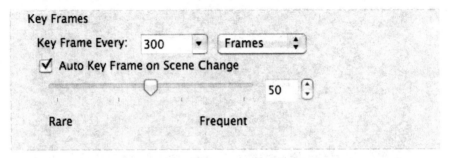

Figure 2-9. You want key frames at scene changes and about every 10 seconds.

I-frames also improve quality when inserted at a scene change, because all subsequent P- and B-frames get a high-quality frame to which to refer. So you also want an I-frame at scene changes, which you enable by checking an eponymous checkbox like that shown in Figure 2-9. Note that if you're using Telestream Episode, the Natural key frame option does the same thing.

The major exception to these general rules is when you're producing for adaptive streaming. As I explain in Chapter 7, the key frame interval used in adaptive streaming is generally shorter and needs to divide evenly into the chunk size for technologies that use chunked files.

In addition, when producing for adaptive streaming, you need key frames at identical locations in all files, so the streams can switch seamlessly among themselves. For this reason, some producers disable the insert key frames at scene changes option, which could result in key frames at different locations, particularly in live events where the streams may be produced by different encoders.

Working With B-frames

Now let's turn our attention to B-frames. As mentioned, the main benefit of B-frames is that they're very efficient from a compression perspective, so they help improve compressed quality. However, files with B-frames are harder to decode because the player has to buffer all reference frames in memory while playing back the file and display them in their proper order.

For this reason, you shouldn't use B-frames when producing video for an iPod or similar device, which are relatively low-powered compared with a computer. Typically, this shouldn't be an issue because if you choose a template for an iPod, the encoding tool won't let you produce a file with B-frames.

Otherwise, when producing for general computer playback, always use B-frames when available. Most encoding tools will let you choose the maximum number of B-frames to insert sequentially between I- and P-frames (or between P- and P-frames), and I recommended inserting three B-frames in sequence.

As you can see in Figure 2-10, which is the B-frame control from Episode Pro, you typically can also set the maximum number of reference frames from which a B-frame can search for and use redundant data. My recommendation here is five frames.

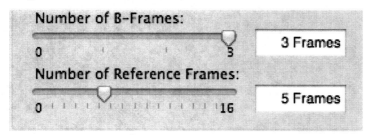

Figure 2-10. Choosing B-frame-related options in Telestream Episode Pro.

What about P-frame-related quantity options? You typically never see these because P-frames are the default. So when producing streaming files, control your I-frame and B-frame options, and P-frames will take care of themselves.

Again, I'll review these options as they related to H.264 in much more detail in Chapter 4; I just wanted to get you acquainted with them early in the process.

Conclusion

OK, now you're hip to most of the key streaming production concepts that you need to go forward, so let's do so! In Chapter 3, you'll learn how H.264 compares with other codecs in terms of quality, playability and other factors, and in Chapter 4, we'll start looking at H.264 encoding parameters.

despread adoption and usage gives H.264 significant momentum going forward. For
le, its inclusion in multiple devices like mobile phones and iPods dramatically reduces
st of chips and other components that enable H.264 playback, creating a natural barrier
y for competitive formats. Overall, though Google's WebM announcement created lots
e in the computer space, it's tough to imagine any codec making significant inroads
264's market share over the next two to five years.

Evolution of H.264 Development and Adoption

Year	ITU International Telecommunications Union (TV, Radio, Phone)	ISO International Standards Organization (Photography, Consumer Electronics, Computers)	Streaming Video Codecs
1984	H.120		
1993	H.261 - video conferencing	MPEG-1	
1994		MPEG-2	
1995	H.263 - better video conferencing		
1999		MPEG-4	QuickTime 4 with MP3
2002	AVC (H.264)	AVC (MPEG-4 Part 10)	QuickTime 6 - MPEG-4
2005			QuickTime 7 - H.264 / First iPod with H.264 playback
2007			Flash - H.264 support
2008			Silverlight - H.264 Support

Figure 3-1. The evolution of H.264 development and adoption.

's audio sidekick is advanced audio coding (AAC), which is designated MPEG-4 Part 3.
H.264 and AAC are technically MPEG-4 codecs—though it's more accurate to call them
eir specific names—and compatible bit streams should conform to the requirements of
4 of the MPEG-4 spec.

64 Wrappers

to say that H.264 is a present that comes in many wrappers (or formats), including
Time, Flash, Silverlight and the H.264 standard wrapper. According to Part 14, MPEG-
; containing both audio and video, including those with H.264/AAC, should use the
extension, while audio-only files should use .m4a and video-only files should use
However, different vendors have adopted a range of extensions that are recognized

Chapter 3: Introduction to H

Before we jump in and start encoding H.264 video, it's useful to know a bit abo
it came from and how it compares with other codecs that you can and should
ing in certain circumstances. You know, kick the tires before you actually buy tl
what we'll do this chapter.

We'll start with a look at what H.264 is and how it has ascended to the top of tl
mid. Then we'll look at the royalty structure associated with H.264, so you can
much H.264 will cost you to use. Then we'll see how H.264 compares with othe
terms of quality and required CPU horsepower to play the files.

All this will give you a good sense of why and how you want to use H.264, and
verting over from other codecs, whether you need to change your encoding p
target platforms. In particular, if you're currently producing with VP6, you'll lea
is the right time to change over to H.264 for Flash deployment. Finally, we'll cc
and VC-1 performance in the Silverlight playback environment, and we'll take
Google's WebM codec to put that into perspective.

What Is H.264?

H.264 is a video compression standard known as MPEG-4 Part 10, or MPEG-4 /
video coding). It's a joint standard promulgated by two international standard
the International Telecommunications Union (ITU) and the International Orga
Standardization/International Electrotechnical Commission (ISO/IEC). Betwee
ISO/IEC, H.264 is the codec of choice for TV, radio and phones, cameras, consu
devices and computers. In terms of streaming, as you can see in Figure 3-1, A
the standards-based bandwagon early and was later followed by both Adobe

H.2
Bo
by
Par

H.

I lik
Qu
4 fi
.mr
.m4

by their proprietary players, such as Apple with .m4p for files using FairPlay Digital Rights Management and .m4r for iPhone ringtones. Adobe has adopted the .f4v extension for H.264 files bound for Flash playback, and H.264-encoded video produced for mobile phones typically uses the .3gp and .3g2 extensions.

Understand that the H.264-specific encoding options that you'll learn in the next chapter apply to all H.264-encoded files, irrespective of the intended wrapper. So though you'll have to make sure that you produce your files in the format compatible with your intended player (which I'll cover in Chapter 5), the H.264-specific encoding parameters themselves remain the same.

Other H.264 Details

Like MPEG-2, H.264 uses three types of frames, meaning that each group of pictures (GOP) is composed of I-, B-, and P-frames, with I-frames like the DCT-based compression used in the DV video format and B-and P-frames referencing redundancies in other frames. We discussed this briefly in Chapter 2, and we will address it in detail in Chapter 4.

Like most video coding standards, H.264 actually standardizes only the "central decoder ... such that every decoder conforming to the standard will produce similar output when given an encoded bit stream that conforms to the constraints of the standard," according to the "Overview of the H.264/AVC Video Coding Standard" published in *IEEE Transactions on Circuits and Systems for Video Technology*. Basically, this means that there's no standardized H.264 encoder. In fact, H.264 encoding vendors can utilize a range of different techniques to optimize video quality, so long as the file plays on the target player. This is one of the key reasons that H.264 encoding interfaces vary so significantly among the various tools.

It's also one of they key reasons that the quality of H.264-encoded files varies so significantly among the different codec vendors. Unlike VP6 and Windows Media, which come from a single vendor and are relatively uniform in terms of quality irrespective of encoding tool, there are multiple H.264 codec developers, and quality varies markedly among them.

H.264 Royalties

H.264 was developed by a consortium of companies who (gasp!) want to get paid for their efforts. To promote this goal, they patented many of the underlying technologies and contracted with a company called MPEG-LA to set up and administrate licensing and collection. Typical customers of MPEG-LA include consumer equipment manufacturers (Blu-ray Disc players and recorders), software developers (encoding programs, streaming players), and content developers.

To explain, companies like Adobe, Apple and Microsoft pay MPEG-LA by the unit to include H.264 decoders in their respective players. There's a maximum annual fee of US $5 million, which is steep but makes including H.264 playback in hundreds of millions of players feasible. Encoding vendors like Sorenson, Telestream and Rhozet are also subject to license fees if their volumes exceed certain limits. Under certain circumstances, content producers who encode with H.264 are also subject to royalties.

MPEG-LA has a licensing FAQ at **www.mpegla.com/main/programs/AVC/Pages/FAQ.aspx**. There's also a PDF file that you can download from that page titled "Summary of AVC/H.264 License Terms." In this section, I'll provide a brief overview of those terms, primarily focused on content producers.

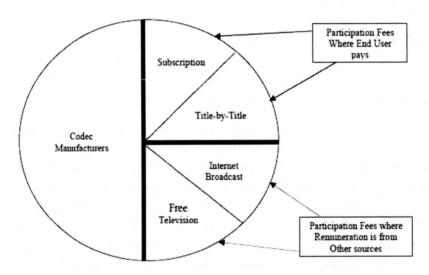

Figure 3-2. H.264 royalty categories per MPEG-LA license summary.

Figure 3-2 presents a chart from the license summary. All those on the left of the wheel are subject to royalties after shipping more than 100,000 units. On the top right, if you're selling H.264-encoded content, on either a subscription or pay-per-view basis, you owe the following, according to the license summary:

Subscription: Royalties start at 100,001 subscribers; for 100,000 and under there is no fee. For 100,000 to 250,000 subscribers, the annual royalty is $25,000, increasing to $50,000 for 250,000 to 500,000 subscribers. Between 500,000 and 1,000,000 subscribers, the annual royalty is $75,000, while services with more than 1,000,000 subscribers owe $100,000 annually.

Title-by-Title (Pay-per-view). There is no royalty for videos that are 12 minutes. For videos longer than 12 minutes, the price is "the lower of 2% of the price paid to the Licensee (on first arms length sale of the video) or $0.02 per title (categories of licensees include legal entities that are (i) replicators of physical media, and (ii) service/content providers (e.g., cable, satellite, video DSL, Internet and mobile) of VOD, PPV and electronic downloads to end users).

The Internet broadcast category is the one that's interesting to most producers. The short answer here is that it's free in perpetuity; so long as you don't charge for your videos, you can use H.264 for free (www.mpegla.com/main/pages/media.aspx).

How it got to be free is mildly interesting. When the H.264 licensing policies were first announced, free Internet video had no royalty until at least January 1, 2011. Then, in February 2010, MPEG-LA announced that royalties would be delayed until December 31, 2015. Then, in August 2010, MPEG-LA announced that its "AVC Patent Portfolio License will continue not to charge royalties for Internet video that is free to end users (known as 'Internet Broadcast AVC Video') during the entire life of this License."

Of course, between February and August, Google launched WebM, the first real competitor to H.264. Though MPEG-LA didn't specify why it waived the royalty, the decision probably related to WebM, at least in part. At a high level, though, any royalty on free Internet video could really muck up the business models for all video-intensive sites, so perhaps this was coming anyway. Either way, the bottom line is that if you're not charging for your Internet video, there is no royalty for encoding with H.264, and there won't ever be.

Comparing H.264 With Other Codecs

We've already seen the compatibility strengths of H.264: It can play on almost any device known to man and in all three streaming architectures. That's swell and all, but how does it compare quality-wise with other technologies, and how hard are the videos to decode? Let's take a quick look.

Looking Backward: VC-1 and VP6

I performed my last H.264/VP6/VC-1 encoding comparison in 2008; since the latter two codecs haven't advanced since then, and H.264 has only gotten better, the results are still valid. You can read and view the results in full at bit.ly/hD9hEo. Here's my pithy summary:

> Overall, H.264 clearly won every trial, from low to high-motion, retaining more detail and showing fewer compression artifacts than either other codec. In general, when the going got tough, VP6 would become slightly blurry, while VC-1 would show more blockiness and other artifacts.

Here's a sample frame that you can see more clearly if you view the article online.

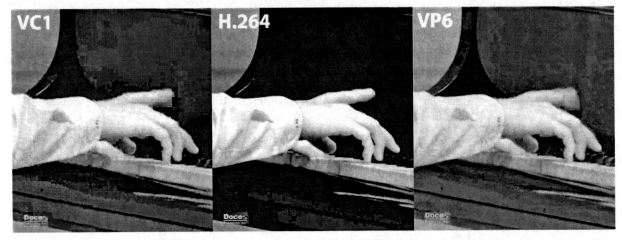

Figure 3-3. H.264 offers better quality than both VC-1 and VP6.

For even more information regarding how these codecs compare quality-wise, check out the handouts from my presentation on the subject at StreamingMedia East 2008 in New York City (bit.ly/hXMQop).

Which Codec Is Hardest to Play Back?

Before Microsoft integrated H.264 playback into Silverlight, it promulgated lots of FUD (fear, uncertainty and doubt) about how hard H.264 was to decode. The basic thrust was that although H.264's quality was good, it wouldn't play smoothly on many computers, so web producers were better off using VC-1 and Silverlight.

To assess the comparative horsepower necessary to play back H.264, VC-1 and VP6, I enlisted the help of both Microsoft and Adobe to create files and players that I could benchmark. You can read about the experience at bit.ly/eWLrEA. Table 3-1 displays the primary results.

As you can see, other than the iMac, upon which neither H.264 nor Silverlight would play smoothly, H.264 required the least CPU horsepower of all the formats. If you want to play the files yourself, here are the URLs:

- Silverlight: www.doceo.com/silverlight/test.html

- H.264: www.doceo.com/HD_Perf_2010/H264_Flash/Main.html

- VP6: www.doceo.com/flv/test.html

- VP6-S: www.doceo.com/flvs/test.html

	Flash VP6-E	Flash VP6-S	Flash H.264	Silverlight
Power Mac, Dual 2.7 GHz PPC G 5,				
Processor (percentage of two CPUs)	72%	66%	86%	108%
Drop frames/audio	No/No	No/No	Yes/Yes	Yes/No
HP xw4100, 3.0 Ghz P4 with HTT				
Processor (percentage of overall CPU)	55%	52%	45%	53%
Drop frames/audio	Yes/Yes	Yes/No	No/NO	No/No
iMac, 2.0 GHz Core 2 Duo				
Processor (percentage of two CPUs)	88%	92%	NA	NA
Drop frames/audio	Yes/Yes	No/No	Yes/Yes	Yes/Yes
HP 8710p, 2.2 GHz Core 2 Duo				
Processor (percentage of overall CPU)	52%	52%	35%	47%
Drop frames/audio	No/No	No/No	No/No	No/No
Dell Precision 390, 2.9 GHz Core 2 Duo				
Processor (percentage of overall CPU)	23%	18%	8%	26%
Drop frames/audio	No/No	No/No	No/No	No/No

Table 3-1. CPU power required to play back these codecs in a browser.

I first ran these tests back in February 2009, and since then, both Adobe and Microsoft have released new versions of their players that can play back H.264 video using the graphics processor on certain graphics chips, which is called "GPU acceleration." So I re-ran a subset of these tests, and I found that H.264's advantage had increased even further, albeit on systems that support GPU acceleration of H.264 video.

	Flash VP6-E	Flash VP6-S	Flash H.264	Silverlight
HP 8710p, 2.2 GHz Core 2 Duo				
Processor (percentage of overall CPU)	46%	50%	24%	43%
Drop frames/audio	No/No	No/No	No/No	No/No
MacBook Pro, 3.06 GHz Core 2 Duo				
Processor (percentage of two CPUs)	38%	39%	23%	36%
Drop frames/audio	No/No	No/No	No/No	No/No

Table 3-2. Updated stats on computers with GPU accelerated H.264 playback.

What About Silverlight?

I've never been a fan of Windows Media/VC-1 quality, and the move to H.264 for Silverlight distribution solves that problem. But what about playback horsepower? To test this, I created two Silverlight pages—one running H.264, the other VC-1—then tested playback on the same two systems. Note that this is a different Silverlight player than the one used in Tables 3-1 and 3-2, which is why the results vary a bit. Table 3-3 shows the results.

	SL-VC1	SL-H.264
HP 8710p, 2.2 GHz Core 2 Duo Processor (percentage of overall CPU) Drop frames/audio	40% No/No	40% No/No
MacBook Pro, 3.06 GHz Core 2 Duo Processor (percentage of two CPUs) Drop frames/audio	40% No/No	44% No/No

Table 3-3. Comparing VC-1 with H.264 playback CPU within Silverlight.

On the Windows computer, the required CPU cycles were the same. On the Mac, H.264 required about 4% more playback CPU, which isn't significant given the major boost in quality. Overall, if you haven't switched over from VC-1 to H.264 for Silverlight because you were concerned that the H.264 streams would be harder to play back, this shouldn't be an issue.

So we know that for Flash and Silverlight, H.264 offers better quality and either equivalent or easier playback than legacy codecs VP6 or VC-1. That's the backward look. What about looking forward to Google's WebM?

Google's WebM

For those who haven't heard of WebM, here's the quick skinny. According to http://www.webmproject.org/about/ is a "royalty-free, media file format designed for the web." Briefly, WebM uses the VP8 video codec that Google purchased from On2 in 2010, the Vorbis audio codec, and a file structure based upon the Matroska container. Though WebM is new, the VP8 codec itself was first launched on September 13, 2008, and comes with some history and some baggage. The history is its predecessor, VP6, which came to prominence when Adobe bundled it into Flash, and is still the most widely used video codec on the Internet today. The baggage is statements in the VP8 press release like:

> With the introduction of On2 VP8, On2 Video now dramatically surpasses the compression performance of all other commercially available formats. For example, leading H.264 implementations require as much as twice the data to deliver the same quality video as On2 VP8 (as measured in objective peak signal to noise ratio (PSNR) testing).

The press release continues:

> In addition, the On2 VP8 bit stream requires fewer processing cycles to decode, so users do not need to have the latest and greatest PC or mobile device to enjoy On2 VP8 video quality.

During the next 11 months, On2 never made VP8 available for testing, at least to me or anyone at *Streaming Media* Magazine, and after the Google signed the agreement to purchase On2 on August 5, 2009, information about VP8 became tougher to find than President Obama's fabled Kenyan birth certificate. Google closed the deal on February 19, 2010, and launched WebM on May 19, 2010.

Google has never repeated the exact claims that On2 made in the initial press release, but when Google makes claims like "highest-quality real-time video delivery," and "low computational footprint," it does draw some skepticism. So the obvious question is, How does WebM's quality and playback compare to H.264?

WebM vs. H.264: Playback CPU

I'm excerpting these results from "WebM vs. H.264: A Closer Look," which I wrote for *StreamingMedia.com* in July 2010 (**bit.ly/hS7jA9**) and later updated for a presentation I gave at StreamingMedia West 2010 in November 2010 (**bit.ly/webmbakeoff**). In terms of playback CPU, here's the performance chart, which compares WebM playback with H.264 Flash and H.264 via HTML5:

	WebM	H.264 - Flash	H.264 - HTML5
MacBook Pro, 3.06 GHz Core 2 Duo			
Processor (percentage of two CPUs)	38%	24%	15%
Differential		-38%	-60%
Acer Aspire One, Intel Atom			
Processor (percentage of overall CPU)	46%	50%	24%
Differential		2%	5%
HP 8710p, 2.2 GHz Core 2 Duo			
Processor (percentage of overall CPU)	38%	39%	23%
Differential		-70%	-47%
iMac, 2 GHz Core 2 Duo			
Processor (percentage of two CPUs)	38%	39%	23%
Differential		2%	-54%

Table 3-4. WebM playback performance compared with H.264 in Flash and HTML5.

On a MacBook Pro with GPU acceleration for H.264 decoding, WebM took 38% of total CPU to play back a 720p file, compared with 24% for H.264 played via Flash, and 15% via HTML5 in Apple Safari. On an Acer Aspire One Netbook without GPU acceleration for H.264, WebM was actually slightly more efficient than H.264 played back either via Flash or HTML5, though the difference wasn't significant. Note that the tests on this small-screen netbook involved a 640x480 file, not 720p.

On an HP 8710w mobile workstation with GPU acceleration for H.264 playback, H.264 via Flash required 70% less CPU power than WebM to play back the 720p file, and H.264 via HTML5 took 47% less CPU power. On my daughter's iMac, WebM and non-accelerated Flash-based H.264 based playback ran neck and neck, while Apple's Safari, presumably with hardware acceleration, proved 54% more efficient than WebM.

Basically, though there are huge swings with the individual browsers, where GPU acceleration exists for H.264, it's significantly more efficient than WebM; where it doesn't, they're neck and neck. At this point, between Flash Player 10.1 with hardware acceleration on supported graphics cards and platforms and Apple's own Safari browser, there are lots of hardware-accelerated H.264 platforms and few if any for WebM, though they will come in time.

Interestingly, on the WebM website, Google says

> Note: The initial developer preview releases of browsers supporting WebM are not yet fully optimized and therefore have a higher computational footprint for screen rendering than we expect for the general releases. The computational efficiencies of WebM are more accurately measured today using the development tools in the VP8 SDKs. Optimizations of the browser implementations are forthcoming.

Truth be told, I'm not that much of a geek, so the low-level development tools were a non-starter for me. However, I did download the DirectShow components to my two Windows computers and played the WebM file via Windows Media Player. On the HP 8710w, CPU load during playback of the same HD WebM file was 18%, with all acceleration disabled, compared with a low of 70% on any of the tested browsers and 21% for hardware-accelerated Flash H.264 playback. On the Acer Aspire One, CPU load dropped to 24%, 30% with hardware acceleration disabled, down from a low of 51% with any of the tested browsers and compared with 53% for non-hardware-accelerated Flash-based H.264 playback.

I'm from Missouri (the "Show Me" state) when it comes to all codec-related claims, so I'm not willing to assume that subsequent updates will reduce browser-based WebM playback loads to these levels. If that occurs, however, the value proposition for WebM as compared with H.264 changes to similar quality, a bit slower encode, but much lower playback requirements—which could be pretty compelling, particularly for low-powered mobile markets.

WebM vs. H.264: Video Quality

To compare the quality of the two codecs, I pitted WebM against the MainConcept and x264 H.264 codecs. You can see a range of still image comparisons and play back the actual test files from links that you can find at bit.ly/x264vh264. Here's a sample image—probably the toughest one in my SD trials—that shows every little difference in quality.

After reviewing this data, I concluded that x264 produced slightly higher-quality than the H.264 files produced by Squeeze with the Main Concept encoder, which was slightly better than WebM. In my view, however, quality differences are irrelevant if the typical viewer wouldn't notice the difference absent side-by-side comparisons at normal data rates. Between H.264 and WebM, a viewer clearly wouldn't notice the difference—the two codecs are at commercial parity when it comes to quality.

Figure 3-4. WebM quality compared with x.264 and MainConcept's H.264.

Overall, as it stands today, WebM's value proposition is basically that it's free and better than Ogg Theora, though it's too tough to decode as compared with the rapidly expanding base of GPU-accelerated H.264. If the browser vendors and Google can reduce CPU playback require- ments to levels shown in Media Player, however, the story changes considerably—at least for pay-per-view and subscription-based video distributors, who currently pay a fee to deliver H.264 video.

Even for those vendors, however, there's that little problem of distributing WebM video to browsers like Internet Explorer 9 and Safari that don't (and won't) support it without a plug-in from Google. Ironically, given the lack of universal HTML5 browser support, the most efficient way to distribute WebM may be via the Flash Player—assuming Adobe adds WebM, and as- suming that the Flash Player's CPU playback requirements are competitive. Of course, if you're distributing to Apple's iDevices, H.264 is your only option, and I suspect it will be your only option far into the future.

If you're publishing free video on the Internet, now that H.264 is royalty-free in perpetuity, there's no financial incentive to switch. If your organization wants to migrate towards HTML5, WebM doesn't provide that single-codec solution. It also is still lower-quality than H.264, however small the difference, and takes longer to encode. Overall, for those not charging for their video, H.264 is still a better solution, and given the rapidly increasing size of the GPU- accelerated installed base, it will likely remain so unless and until Google creates distribution channels that you can't access with H.264. In short, you can ignore WebM for the next 18 to 24 months or so—say till the end of 2013.

For those interested in more information on WebM, check out "VP8/WebM—A Collection of Resources," which you can find at **bit.ly/webmresources**. In particular, take a look at the in- depth technical analysis of VP8 from *Diary of an x264 Developer* that you can find at x264dev.multimedia.cx/?p=377, noting the potential for patent issues that may hinder WebM's adoption.

Conclusion

OK. So you've learned a bunch of background information about H.264 and how it compares with relevant codecs, old and new. Enough background; in Chapter 4, you get to start producing some H.264 videos.

Chapter 4: H.264 Encoding Parameters

OK, now you know the basics, let's review a couple of items and then jump into the most common H.264 encoding parameters.

First, while we'll learn the most commonly applied H.264 encoding parameters in this chapter, not all encoding tools expose these options to you. This isn't necessarily a bad thing; sometimes less is more. To make this point, I'll incorporate H.264 encoding screens from the most popular encoding tools into this discussion.

Second, not all encoding tools can produce all files in the desired wrapper; for example, Apple Compressor can produce MOV and MP4 files, but not F4V. I'll discuss the various wrappers in detail in Chapter 5, and I'll also detail when and how you can change a file extension without ruining the file.

Basic H.264 Encoding Parameters

Let's start with the basics, profiles and levels.

Profiles and Levels

Profiles and levels are the most basic H.264 encoding parameters, and are available in one form or another in most H.264 encoding tools. According to *Wikipedia* (en.wikipedia.org/wiki/H264), a profile "defines a set of coding tools or algorithms that can be used in generating a conforming bit stream," whereas a level "places constraints on certain key parameters of the bit stream." In other words, a profile defines specific encoding techniques that you can or can't use when encoding a file (such as B-frames), while the level defines details such as the maximum resolutions and data rates within each profile.

	Baseline	Main	High
I and P Slices	Yes	Yes	Yes
B Slices	No	Yes	Yes
Multiple Reference Frames	Yes	Yes	Yes
In-Loop Deblocking Filter	Yes	Yes	Yes
CAVLC Entropy Coding	Yes	Yes	Yes
CABAC Entropy Coding	No	Yes	Yes
Interlaced Coding (PicAFF, MBAFF)	No	Yes	Yes
8x8 vs. 4x4 Transform Adaptivity	No	No	Yes
Quantization Scaling Matrices	No	No	Yes
Separate Cb and Cr QP control	No	No	Yes
Separate Color Plane Coding	No	No	No
Predictive Lossless Coding	No	No	No

Figure 4-1. Encoding techniques enabled by profile, from Wikipedia.

Take a look at Figure 4-1, which is a screen grab from a features table from *Wikipedia's* description of H.264. On top are H.264 profiles, including the Baseline, Main and High, which are the profiles most frequently supported in computer- and device-oriented players. On the left are the different encoding techniques available, with the table detailing which are supported by the respective profiles.

As you would guess, the higher-level profiles use more advanced encoding algorithms and produce better quality. This is shown in Figure 4-2.

To produce this figure, I encoded the same source file to the same encoding parameters (720p@1200 kbps). The file on the left used the High profile with maximum quality settings enabled, and Baseline was used on the right. A check of the chart in Figure 4-1 reveals that the High profile enables B slices (usually called B-frames), and the higher-quality CABAC encoding, defined below, as well as some other more technical parameters. As you can see, these do help the High profile deliver noticeably higher-quality video than the Baseline profile.

So the High (and Main) profiles deliver better quality than the Baseline profile; what's the catch? The catch is, as you use more advanced encoding techniques, the file becomes more difficult to decompress and may not play smoothly on older, slower computers or low-powered devices.

Figure 4-2. A file encoded using the High profile (on the left) retained much more quality than a file encoded using the Baseline profile (on the right).

As a brief aside, this illustrates one of the two trade-offs typically presented by H.264 encoding parameters: better quality for a harder-to-decompress file. The other trade-off is a parameter that delivers better quality at the expense of encoding time. In some rare instances, as with the decision to include B-frames in the stream, you trigger both trade-offs, increasing both decoding complexity and encoding time.

OK, back to profiles. At a high level, think about profiles as a convenient point of agreement for device manufacturers and video producers. Mobile phone vendor A wants to build a phone that can play H.264 video, but needs to keep the cost, heat and size requirements down. So the crafty chief of engineering searches and finds the optimal processor that's powerful enough to play H.264 files produced to the Baseline profile. If you're a video producer seeking to create video for that device, you know that if you encode using the Baseline profile, the video will play.

Accordingly, when producing H.264 video, the general rule is to use the maximum profile supported by the target playback platform, since that delivers the best quality at any given data rate. If you're producing for devices, this typically means the Baseline profile, but check the documentation for that device to be sure. For example, all pre-5G iPods and iPhones can only play the Baseline profile, while the iPad can play video encoded with the Main profile, as can 5G iPhone/iPod touch devices. If you're producing for Flash consumption on Windows or Mac computers, always use the High profile.

This sounds nice and tidy, but understand this. While encoding using the Baseline profile ensures smooth playback on your target mobile device, using the High profile for files bound for computer playback doesn't provide the same assurance. That's because the High profile supports H.264 video produced at a maximum resolution of 4096x2048 and a data rate of 720 Mbps. Few desktop computers could display a complete frame, much less play back that stream at 30 fps.

Accordingly, while producing for devices is all about profile and level, producing for computers is all about your video configuration. Here, as we saw in Chapter 3, the general rule is that decoding H.264 video is about as computationally intense as decoding VP6, or Windows Media, for that matter. So long as you produce your H.264 video at a similar resolution and data rate as the other two codecs, it should play fine on the same class of computer using the High profile.

In general, this means that SD video produced at 640x480 resolution and below should play fine on most post-2003 computers. If you're producing at 720p or higher, these streams won't play smoothly on many of these computers, and you should consider offering an alternative SD stream for these viewers. You'll learn how to customize your configuration for your target computers in Chapter 5.

In summary, here's the skinny on profiles:

- Use the High profile when producing for computers.
- Check the specs on all devices (iPod/iPhone are Baseline, iPad is Main).

H.264 Levels

What about H.264 levels? Again, as mentioned above, levels provide bit rate, frame rate and resolution constraints within the different profiles, which is shown in Figure 4-3.

In this role, levels enable primarily device vendors to further specify the types of streams that will play on their devices. For example, the Apple iPad will play video encoded with the Main profile, but only up to level 3.1. As with profiles, if you encoded to parameters beyond the specified level, either the file won't load on the device (iTunes typically won't load non-conforming files on any iDevice) or it won't play.

Accordingly, when you're producing for devices, you need to ensure that your encoding parameters don't exceed the specified level, which again should be designated by the device manufacturer. In contrast, levels are irrelevant when encoding for computer playback because all three streaming players—QuickTime, Flash and Silverlight—can play any level of the Baseline, Main or High profiles. So you can't choose the wrong level—the player will always attempt to play the file anyway.

Level number	Max video bit rate (VCL) for Baseline, Extended and Main Profiles	Max video bit rate (VCL) for High Profile	Examples for high resolution @ frame rate (max stored frames) in Level
1	64 kbit/s	80 kbit/s	128x96@30.9 (8) 176x144@15.0 (4)
1b	128 kbit/s	160 kbit/s	128x96@30.9 (8) 176x144@15.0 (4)
1.1	192 kbit/s	240 kbit/s	176x144@30.3 (9) 320x240@10.0 (3) 352x288@7.5 (2)
1.2	384 kbit/s	480 kbit/s	320x240@20.0 (7) 352x288@15.2 (6)

Figure 4-3. Levels constrain bit rate, frame rate and resolution for the different profiles. From Wikipedia.

Many encoding tools don't let you specify a level. When you're producing for devices with these encoders, be sure to use the supplied device templates to ensure that the resolution, frame rate and data rate don't exceed those supported in the specified level. Other encoding tools do specify the level, and present an error message if you encode using parameters that exceed the selected level.

This is shown in Figure 4-4, a screen shot of Adobe Media Encoder. If you were producing for a device with this encoder, compatibility with the selected level is key, so you'd have to dial back your encoding parameters to the selected level. If you were producing for computer playback, you would simply choose a higher level that lets you produce the file at the selected parameters, and restart encoding.

Figure 4-4. Oops, the selected frame size exceeds the parameters allowed for the selected level.

Again, however, just because the Flash Player won't refuse to play a file encoded using the High profile doesn't mean that it will play that file smoothly. While profiles and levels are critical for devices, choosing the right resolution and data rate for your streaming video file is much more important when targeting computer playback. I cover this in detail in Chapter 5.

In summary, here's the skinny on levels:

- Critical when encoding for devices.

- Not so important when encoding for computers, where the resolution of your video is the most important consideration.

Apple Compressor and Adobe Media Encoder

Let's take a break from theory for a quick gaze at reality, in the form of the Apple Compressor encoding interface shown in Figure 4-5. As you can see, there's no mention of profiles or levels whatsoever.

Figure 4-5. Apple Compressor's H.264 encoding parameters.

The key option in this interface is Frame Reordering, which I've circled. Check this, and Compressor produces the file using the Main profile with B-frames, with a B-frame interval of one and two reference frames, for an IBPBPBPBPBPB ... cadence. Leave it unchecked, and Compressor produces the file using the Main profile without B-frames. To produce a file using the Baseline profile, you have to use an iDevice preset that has a completely different interface. At the other end of the spectrum, at this point in time, Compressor can't produce files using the High profile.

What about Adobe Media Encoder? The encoding parameters are a little more straight-forward, but equally austere, with configuration options for Profiles and Levels, but nothing else. Looking ahead to Figure 4-9 and beyond, you'll see that other tools provide a much greater range of H.264 encoding options. OK, fun's over; back to theory.

Entropy Coding

When you select the Main or High profiles, some encoding tools provide two options for Entropy Coding Mode: CAVLC, which stands context-based adaptive variable length coding, and CABAC, which stands for context-based adaptive binary arithmetic coding.

Of the two, CAVLC is the lower-quality, easier-to-decode option, while CABAC is the higher-quality, harder-to-decode option. This is shown in Figure 4-6 from Rhozet Carbon Coder.

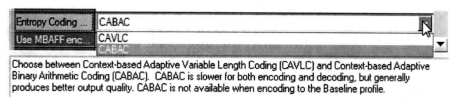

Figure 4-6. Your Entropy Coding choices: CABAC and CAVLC.

Though results are source-dependent, CABAC is generally regarded as being between 5-15% more efficient than CAVLC. This means that CABAC should deliver equivalent quality at a data rate of between 5 and 15% lower, or better quality at the same data rate.

CABAC   CAVLC

Figure 4-7. A 720p file produced using CABAC on the left, CAVLC on the right.

In my own tests, CABAC produced noticeably better quality, though only in HD test clips encoded to very low data rates. This is shown in Figure 4-7, which shows a 720p file produced with CABAC on the left and CAVLC on the right, both to the same 800 kbps video data rate. Though neither image will win an award for clarity, the ballerina's face and other detail is clearly more visible on the left. The bottom line is that CABAC should deliver better quality, however modest the difference. Now the question is, How much harder is the file to decompress and play?

	CABAC	CAVLC	Difference
HP 8710p, 2.2 GHz Core 2 Duo	31%	31%	0.6%
Power Mac, Dual 2.7 GHz PPC G 5	36%	34%	1.8%

Table 4-1. CPU consumed when playing back H.264 files encoded using CABAC and CAVLC.

It turns out, not that much. I tested this on two of the less powerful multiple-core computers in my office, one a HP notebook with a Core 2 Duo processor, and the other a Power PPC-based Power Mac. As you can see in Table 4-1, the CABAC file increased the CPU load by less than 1% on the HP notebook, and less than 2% on the Mac. Based on the improved quality and minimal difference in the required playback CPU, I recommend choosing CABAC whenever the option is available.

What Would YouTube Do?

As you likely know, YouTube streams about 70% of the video streams seen on the Internet, as most of us with teenagers can likely attest. I sure can. So when it comes to producing H.264 video, it's instructive to see how YouTube re-encodes videos that are uploaded.

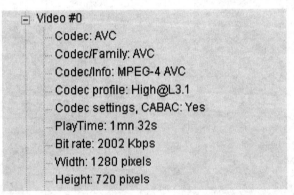

Figure 4-8. A MediaInfo analysis of a 720p YouTube file.

To do this yourself, download a 720p YouTube file; I use the free Firefox plug-in Download Helper (**www.downloadhelper.net**) for the job. Then download free file analysis tool MediaInfo (**www.mediainfo.sourceforge.net**), install it and analyze the downloaded YouTube file. You should see something that approximates the screen shot in Figure 4-8, showing a file encoded with the High profile with CABAC enabled, confirming my recommendations.

Squeeze and Episode

Just to take the intimidation factor out of these parameters, let's take a quick look at how they're presented in two of the most popular batch-encoding tools on the market. On the left in Figure 4-9, you can see Sorenson Squeeze, with clearly labeled options for AVC Profile and Entropy Coding Mode, with a similar presentation for Telestream Episode on the right. Your versions of the programs may look slightly different, but now that you know what the options mean, you should be able to figure them out in any program that enables them.

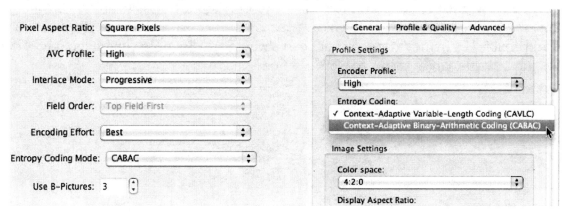

Figure 4-9. Choosing the profile and entropy coding technique in Sorenson Squeeze
(on the left) and Telestream Episode.

OK, these are the basic H.264 parameters; let's move to more advanced controls.

I-, B- and P-frame Controls

We looked at these in Chapter 2, but let's quickly review. I-frames, also known as key frames, are completely self-referential and don't use information from any other frames. I-frames are larger than B-frames or P-frames, and are the highest-quality of the three, but the least efficient from a compression perspective. We learned that in most instances, you should use a key frame interval of 10 seconds, and insert key frames at scene changes.

P-frames are "predicted" frames. When producing a P-frame, the encoder can look backward to previous I- or P-frames for redundant picture information. P-frames are more efficient than I-frames, but less efficient than B-frames.

B-frames are "bi-directional predicted" frames. When producing B-frames, the encoder can look both forward and backward for redundant picture information. This makes B-frames the most efficient frame of the three, but also the most difficult to decode, so they are not available when producing using H.264's Baseline profile. Our default interval for B-frames was three, which would produce a stream configured like this:

IBBBPBBBPBBBPBBBPBBBPBBB ...

and so on until the next I-frame.

Those are the basics; let's look at some H.264-specific I- and B-frame-related controls.

IDR Frames

Figure 4-10 shows the key frame controls from Sorenson Squeeze. Again, unless you're producing for adaptive streaming (in which case, see Chapter 7), use an interval of around 10 seconds and enable key frames at scene changes. I always leave Squeeze's Rare/Frequent slider at the default 50 value, which seems to work just fine.

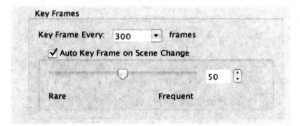

Figure 4-10. Sorenson Squeeze's key frame settings.

One other I-frame configuration option that you'll see in Telestream Episode and some higher-end encoders is the IDR frame (Figure 4-11). Briefly, the H.264-specification enables two types of I-frames: normal I-frames and IDR frames. With IDR frames, no frame after it can refer back to any frame before it. In contrast, with regular I-frames, B- and P-frames located after the I-frame can refer back to reference frames located before it.

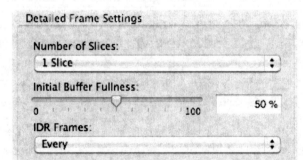

Figure 4-11. Telestream Episode's IDR settings. I discuss the concept of Slices below.

In terms of random access within the video stream, playback can always start on an IDR frame because no frame refers to any frames behind it. However, playback cannot always start on a non-IDR I-frame because subsequent frames may reference previous frames.

Since one of the key reasons to insert I-frames into your video is to enable interactivity, I use the default setting of 1, which makes every I-frame an IDR frame. If you use a setting of 0, only the first I-frame in the video file will be an IDR frame, which could make the file sluggish during random access. A setting of 2 makes every second I-frame an IDR frame, while a setting of 3 makes every third I-frame an IDR frame, and so on.

Here's a blurb from the Telestream Episode help file:

> An IDR frame is an I-frame whose preceding frames cannot be used by predictive frames. More distant IDR frames may allow more efficient compression but limits the ability of a player to move to arbitrary points in the video. In particular, QuickTime Player may show image artifacts when you scrub the timeline unless every I-frame is an IDR frame.

Again, when this option is available, I always make every I-frame an IDR frame.

Working with B-frames

Let's take B-frames to the next level, sticking with Sorenson Squeeze for a moment. The first question is whether B-frames improve the quality of your video—after all, we know that they make the file tougher to decode, so why bother to use them if they don't make your video look better? In a phrase, the answer is they do, as you can see in Figure 4-12.

I like using this frame as an example because it combines the end of a high-motion sequence with lots of detail. On the left, in the file produced with B-frames (setting of three, of course), you can see a lot more of that detail than you can in the file on the right, which was produced without B-frames. For this reason, you should use B-frames whenever they are allowed on the platform that you're targeting. If you're encoding with the Main or High profiles, use B-frames.

Figure 4-12. A file encoded with B-frames on the left, without them on the right.

You get the B-frame interval of three by now, as well as five reference frames, so let's look at other B-frame parameters presented in Figure 4-13, another screen shot from Sorenson Squeeze.

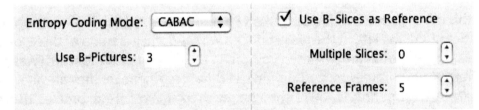

Figure 4-13. Squeeze's B-frame settings.

On the left, we see Use B-Pictures set to 3, which is simple enough (B-Pictures are the same as B-frames in Sorenson-speak). The new configuration options are on the right. Question 1 is whether to "Use B-Slices as Reference." I'll get to the Slice vs. Frame discussion in a moment.

To explain this control, by default, no frames attempt to find redundancies from B-frames during encoding; I-frames stand alone, P-frames look for redundancies with I- and other P-frames, while B-frames also look to I- and P-frames. The reason for this relates back to MPEG-2 compression, where the thought was that since B-frames are the lowest-quality frames, grabbing referential information from those frames would only spread the poor quality. With H.264, B-frames can have very high quality, so this default as applied to H.264 is somewhat vestigial.

If you click the Use B-Slices as Reference checkbox, P- and B-frames can refer to B-frames during encoding. Should you? While enabling this function may improve quality at a given file size to some degree, it also makes the video harder to play back and can cause some risk of incompatibility on hardware devices like the iPad or set-top boxes. If you're trying to wring the last bit of quality out of your files, enable it; if compatibility is more important, disable it.

Question 2 is, What the heck are slices? Briefly, when you use multiple slices, the encoder divides each frame into multiple regions and searches for redundancies in other frames only within the respective region. This can accelerate encoding on multi-core computers, because the encoder can assign the regions to different cores. However, since redundant information may have moved to a different region between frames, say in a panning or tilting motion, encoding with multiple slices may miss some redundancies, decreasing the overall quality of the video.

Here's a snippet from the Telestream Episode help file:

> Speed up processing by transcoding parts, slices, of the same frame in parallel. Using more slices may decrease image quality somewhat as redundancies between parts of the frame cannot be fully utilized.

This is illustrated in Figure 4-14. If this video were encoded in a single slice, movement within the quadrants would be irrelevant, and the encoder could find inter-frame redundancies irrespective of where the content is (or was) in the frame. If that distinguished rider on the left (then Governor, now Senator Mark Warner from Virginia) started on the left and moved through the video toward the right in a sequence of five or six frames, there would be significant inter-frame redundancies, which translates to very good quality.

Figure 4-14. This video frame cut into four slices.

With slices, however, the encoder can't refer to information in previous or subsequent frames that wasn't in that slice, so as the rider moves from left to right, significant inter-frame redundancies may go unrealized. For this reason, unless you're in a significant hurry, I recommend setting slices to the lowest value, which is 0 for Squeeze and 1 for Episode.

What about the frame vs. slice nomenclature? Well, a slice is obviously a region within a frame. If you selected 4 slices, when Squeeze asks whether you want to Use B-Slices as Reference, each slice within a B-frame would potentially be a reference for that slice in another B- or P-frame. If you have 0 slices, the frame is the same as a slice, so you can use the term interchangeably. Slices aren't really about B-frames, but since Squeeze presented the Multiple Slices option within the other B-frame settings, it made sense to discuss it here.

Adaptive B-frames

Let's have a look at Episode's B-frame-related controls, as shown in Figure 4-15. Most should appear pretty familiar by now, including Natural key frames for key frames at scene changes, the key frame distance of 300 frames, with three B-frames and five reference frames. The control that you haven't seen is the checkbox to "Use adaptive B-frames."

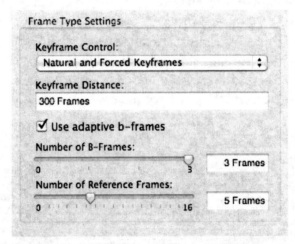

Figure 4-15. Episode's key frame and B-frame-related controls.

Briefly, adaptive B-frame placement allows the encoder to override the B-frame setting when it will enhance the quality of the encoded stream—for instance, when it detects a scene change and substitutes an I-frame for the B-frame. For single-stream encoding, I enable this setting. When encoding for adaptive streaming, I disable this (and use only Forced key frames).

Search-Related Settings

Figure 4-16 shows the H.264 encoding parameters from Rhozet Carbon Coder, which is one of my favorite high-end encoding tools. I include this not to induce drowsiness, which I'm sure I did about 30 minutes ago, but to illustrate that there are multiple search-related settings that encoding tools can deploy. The trade-off with these types of configuration options is encoding time and quality. For example, if you check the three "Use fast" check boxes, encoding speed will increase, but you might be leaving some quality on the table.

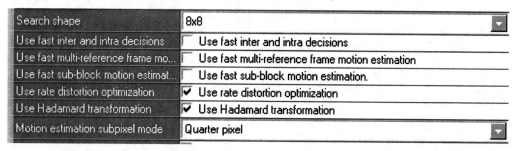

Figure 4-16. Search-related H.264 encoding parameters in Rhozet Carbon Coder.

Carbon Coder is a $5,000 tool, and users expect lots of configurability for that price. In the sub-$1,000 class of encoding tools, like Squeeze and Episode, the program designers have decided to abstract these individual configuration options, and other search-related options, into a single control. For example, with Squeeze, you get to choose Best, Medium or Fast in the Encoding Effort drop-down list.

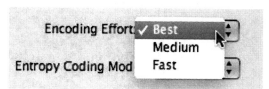

Figure 4-17. Squeeze's Encoding Effort controls.

Best is the default setting, and in my tests, the quality difference between files encoded with Fast and Best selected was noticeable, but not striking. You can see for yourself in Figure 4-18.

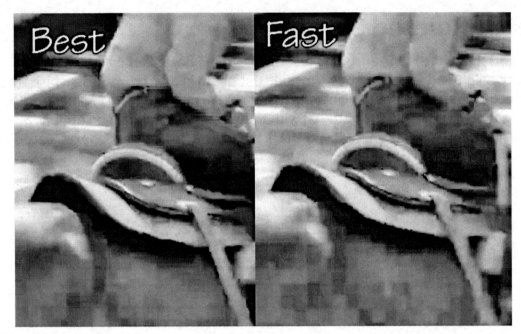

Figure 4-18. In this high-motion sequence, Squeeze's Fast option had more noticeable blocks than Best.

On the other hand, for a 3:10 (min:sec) 720p test file, encoding in Best mode took 18:37 on my Mac Pro, which has two 2.93 GHz quad-core Intel Nehalem Xeon CPUs with 18 GB of 1067 MHz DDR3 RAM. In contrast, with Fast selected, the file encoded in 8:10, a drop of more than 50%. File sizes were almost identical. Certainly, if you're in a hurry to get a draft file out to a client, go with Fast mode; though definitely go with Best for final production.

With Episode, you get to choose a point on a slider from 10% to 100%, with 10% being the low-quality, fast option and 100% the slower, high-quality option. Though the default is 90%, the Episode help file reads:

> 10 represents the fastest encoding, with most of the advanced features turned off, 100 represents the most advanced coding mode, yielding the best quality, but also taking a considerably longer time. In general, values over 50 yield very small improvements in visible image quality.

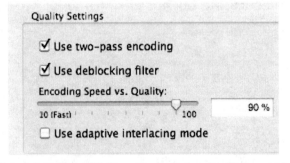

Figure 4-19. Episode's Encoding Speed vs. Quality controls.

My tests confirmed those results, as video quality at 50% and 100% was virtually identical. In terms of encoding speed, however, at 50% quality, the 3:10 (min:sec) 720p test file encoded in 3:33, which ballooned up to 8:03 at 100%. If encoding time is an issue, you should run your own trials at 50% and 100% and gauge the difference in quality and encoding time. If your results are like mine, you could save a boatload of time encoding at the 50% setting and not even notice any quality loss. Interestingly, at 10% quality, the file took 3:20 to encode—only about 13 seconds faster—but the quality was abysmal.

Working With Rhozet Carbon Coder

We've addressed the abstracted search parameters used by Squeeze and Episode; now, let's get down and dirty with Rhozet Carbon Coder, looking at the basic and advanced H.264 encoding controls. The basic controls, shown in Figure 4-20, should be old hat by now.

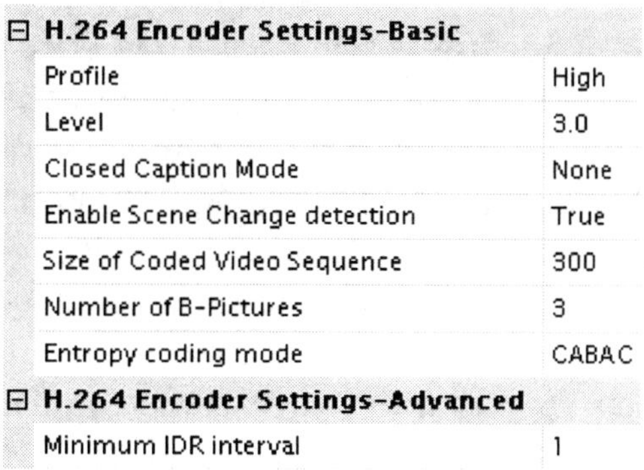

H.264 Encoder Settings–Basic	
Profile	High
Level	3.0
Closed Caption Mode	None
Enable Scene Change detection	True
Size of Coded Video Sequence	300
Number of B-Pictures	3
Entropy coding mode	CABAC
H.264 Encoder Settings–Advanced	
Minimum IDR interval	1

Figure 4-20. Rhozet Carbon Coder's Basic H.264 encoding settings.

Profile and Level are straightforward, as is the Enable Scene Change detection checkbox. Size of Coded Video Sequence is the key frame interval, and 300 is the recommended setting, along with a B-frame interval of three and CABAC for entropy encoding mode. We've covered IDR interval, and 1 is both the default and the recommended setting. Minimum IDR interval defines the minimum number of frames in a group of pictures, and 1 is also the recommended setting for all but the most dynamic footage.

For example, in a very dynamic MTV-like sequence, a minimum IDR interval of 1 could result in frequent I-frames, which could degrade overall video quality. For these types of videos, you could experiment with extending the minimum IDR interval to 30 to 60 frames, to see if this improves quality. For most videos, however, the default interval of 1 provides the encoder with the necessary flexibility to insert frequent I-frames in short, highly dynamic periods, like an opening or closing logo, while using the designated interval of 300 for most footage.

Use adaptive B-frame placement	True
Reference B-Pictures	False
Allow pyramid B-frame coding	False
Reference frames	5
Number of Slices per picture	1

Figure 4-21. Carbon Coder's B-frame options.

Figure 4-21 contains Carbon Coder's B-frame-related controls. As you recall, enabling adaptive B-frame placement allows the encoder to override the Number of B-Pictures value when it will enhance the quality of the encoded stream—for instance, when it detects a scene change and substitutes an I-frame for the B. I always enable this setting for single-file encodes, and disable for adaptive streaming.

Reference B-Pictures lets the encoder use B-frames as a reference frame for P-frames, while Allow pyramid B-frame coding lets the encoder use B-frames as references for other B-frames. I typically don't enable these options because the quality difference is negligible, and these options can cause playback to become unstable in some environments. Note that Rhozet's helpful white paper "Advanced H.264 Encoding with Carbon Coder," which you can download at www.rhozet.com/products_whitepapers.html, disagrees, and recommends enabling both Reference B-Pictures and pyramid B-frame coding in all encodes. Otherwise, our respective recommendations are in synch. We end with a couple of easy ones: reference frames at five and number of slices at one.

Finally, Figure 4-22 contains Carbon Coder's search-related options. First is search shape, which can be either 16x16 or 8x8. The latter (8x8) is the higher-quality option, with the trade-off being longer encoding time. The next three "fast" options allow you to speed encoding time at the possible cost of quality. I typically disable these options.

Search shape	8x8
Use fast inter and intra decisions	False
Use fast multi-reference frame motion estimation	False
Use fast sub-block motion estimation.	False
Use rate distortion optimization	True
Adaptive Quantization Mode	None
Adaptive Quantization Strength	0
Use Hadamard transformation	True
Motion estimation subpixel mode	Quarter pixel

Figure 4-22. Rhozet Carbon Coder's search-related H.264 encoding settings.

Adaptive Quantization Mode and Adaptive Quantization Strength are advanced settings that reallocate bits of data within a frame using one of the three selected criteria: brightness, contrast or complexity. I would only experiment with these settings when areas in the video frame are noticeably blocky. Unfortunately, operation is extremely content-specific, which makes it impossible to offer general advice regarding which techniques and values to use.

Both the rate distortion optimization and Hadamard transformation settings can improve quality but lengthen encoding time; I usually enable both. Finally, the Motion estimation sub-pixel mode defines the granularity of the search for redundancies, with Quarter pixel representing the highest-quality option, though the slowest to encode. Full pixel is the fastest but lowest-quality. Again, in my low-volume production environment, I always use the Quarter pixel option.

Conclusion

For most producers, the foregoing is all that you'll ever need to know about configuring your H.264 video because most encoding tools simply won't expose more controls. The primary exception are those producers who experiment with the x264 codec, which is quite capable and is typically deployed in an interface that exposes dozens more configuration options than I've explained here. The only problem is that the documentation of these options, though relatively easy to find on the Internet, seldom ties the individual options to a usage case.

We also know that many advanced encoding parameters come with side effects like much longer compression times or slower playback on some target machines.

In my view, sometimes simpler is better. Though there are compressionists who produce amazing results by massaging these configuration options to the max, it's very challenging to derive one set of options that works exceptionally well with a broad range of source footage. Experimenting with these options can also be very time-consuming. If you're encoding the next *Star Wars* movie and getting paid by the hour, by all means, invest the time to identify the absolute best parameters. On the other hand, if you're in a high-volume production environment with lots of disparate content, find a set of parameters that produces at least good quality for all of your video footage and go tackle another problem.

But now you know the basics, and certainly enough to use 95% of the commercial encoding programs out there. That's only half the battle though, because unless you know how to choose basic configuration options including resolution, data rate and the like, you could create a very high-quality file that simply won't play on many of your target computers. So in the next two chapters, we apply what we've learned here to determine how to configure our H.264 video for general-purpose computers (Chapter 5) and also for devices like the Apple iPod and iPad (Chapter 6).

Chapter 5: Producing H.264 for Computer Playback

Last chapter, you learned the basic H.264 configuration options; in this chapter, you'll learn how to configure your video for playback on a computer, whether for Flash, QuickTime, Silverlight or HTML5. This chapter has five major components. You'll start by reviewing some configuration options learned during the first couple of chapters to make sure we're all on the same page regarding factors such as progressive vs. interlaced, VBR vs. CBR and the like.

Then, we'll learn the configuration requirements for playback in Flash, QuickTime, Silverlight and HTML5—more specifically, the file naming conventions and format requirements for playback in each environment. With these basics behind us, we'll ask, "How the heck do I pick a resolution and data rate for my video files?"

To answer this, we'll look at three factors. First, we'll examine the results of playback tests that detail how well H.264 videos play on multiple computer platforms. Next, we'll look the configurations used by major corporate and broadcast Internet sites, from Apple to YouTube, with stops in between at CBS, Starbucks and many others. These sites distribute millions of streams daily, and their configurations tell us a lot about the bandwidth their viewers have and the configuration of videos that their viewers' computers can easily play.

Finally, we'll look at the importance of the bits-per-pixel calculation, and use this to produce specific data rate recommendations for the most common video resolutions used on the internet, including 4:3 and 16:9 aspect ratios, at 24, 25 and 29.97 fps. Not to date myself, but it will all come together like an *A-Team* episode, with the bad guys vanquished, the hero getting the girl and your video playing like a charm for millions of remote viewers around the globe.

Preliminaries: What We Already Know

Let's start with what we already know as a review, using Figure 5-1, a screen shot from Sorenson Squeeze, as a guide, starting on the top left. In Chapter 2, we learned about constant and variable bit rate encoding, along with single, two-pass and multi-pass. Though it takes longer to run multi-pass encoding, I'll use that with Squeeze because my encodings aren't time-critical. We also covered frame rate in Chapter 2; I'll go 1:1, which means one encoded frame for each frame in the source video, so if your input is 29.97 fps, your output will be 29.97 fps.

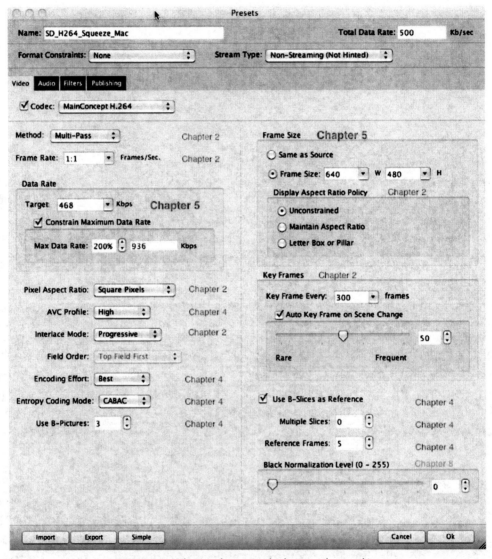

Figure 5-1. What we know and where we learned it.

We learned what data rate is in Chapter 1. In this chapter, we'll learn how to choose the data rate for our files. Moving further south, we learned about choosing the pixel aspect ratio and interlace mode in Chapter 2, and about H.264 encoding parameters like profile and entropy coding mode in Chapter 4.

On the upper right, we see that frame size is the other major decision that we'll learn to make in this chapter. We covered display aspect ratio and key frames in Chapter 2, and about additional H.264 encoding parameters like B-slices and reference frames in Chapter 4.

So hopefully this screen shot makes it clear that by now, you should know pretty much every configuration option save data rate and resolution, which we'll learn in this chapter. Got it? Good. Let's move to producing H.264 for the various playback environments.

Producing H.264 for Specific Environments

As we've learned, H.264 can come in a MP4 file, an F4V file, an FLV file, a WMV file, an M4V file, an MPG file, a 3GP file and even an M2T or MTS file. Fortunately, the H.264-specific parameters that you learned in Chapter 4 are the same when producing for all these wrappers, but you gotta know which wrapper to use for your target playback environment.

To be more specific, by "wrapper," I mean container format, or the format into which the file is written. Typically, you identify container format by the file extension, with MP4 meaning the standard MPEG-4 container format, MOV meaning the QuickTime format and so on. As you'll learn, Flash and Silverlight are very flexible regarding the container format, and will play virtually any H.264-encoded file. QuickTime is bit more persnickety, and devices like the iPhone even more so.

Let's start at the beginning, with a quick review that I'll supplement with more format-specific data in the individual sections below. H.264 is a component of the MPEG-4 specification, and MPEG-4 Part 14 defines the media format for MPEG-4 files, and yes, it was based upon Apple's QuickTime. Under Part 14, files with an .mp4 extension (video.mp4) contain video or audio and video, while files with an .m4a contain just audio. You might call these files as produced in the MP4 wrapper, or MPEG-4 container format.

Apple was among the first to use the H.264 codec, encapsulating it in a MOV or MP4 file for desktop playback via QuickTime, which can play files produced with either extension, and as M4V files when producing for Apple devices like the iPod/iPhone/iPad. Apple uses the .m4a extension for audio-only H.264 files (encoded with AAC audio). Note that you don't have to use the .m4v extension with iDevices; MOV and MP4 files will play just fine.

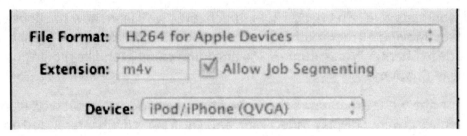

Figure 5-2. Apple Compressor creating an .m4v file for playback on Apple Devices.

Adobe was next to adopt H.264, and it designed the Flash Player to play H.264 files encoded into the MOV, MP4 and 3GP formats. The Flash Player doesn't care about extensions; it looks into the file itself to determine the format and whether it's compatible. In the past, .flv was the default extension for Flash files with the VP6 codec; as you can see in Figure 5-3, a screen from the Adobe Media Encoder, Adobe created the .f4v extension to designate files produced for Flash playback using the H.264 codec.

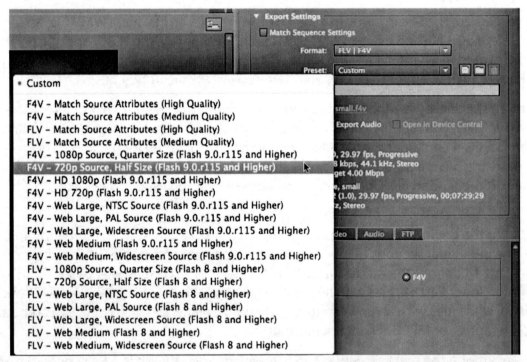

Figure 5-3. Adobe Media Encoder uses the .f4v extension when producing files with the H.264 video codec.

Next into the H.264 sandbox was Microsoft, which did not create its own extension, but vowed to support most commonly used H.264 container formats, including MP4, FV4, 3GP and MOV. Finally, three HTML5-compatible browsers currently play H.264 video—Apple Safari, Google Chrome, and Microsoft Internet Explorer 9—though in early 2011, Google indicated that it will remove H.264 playback from subsequent releases of Chrome. The preferred container format for HTML5 is the standard MP4 wrapper, since the .mov extension might call the QuickTime player, and .f4v the Flash Player.

Before getting to environment-specific configuration options, let's look at one issue that will likely affect playback performance irrespective of whether you're using Flash, QuickTime, Silverlight or HTML5. Specifically, the moov atom issue.

Where Dat Moov Atom?

Have you ever posted a Flash or QuickTime video on the Net for progressive playback and noticed that it wouldn't start to play until it was fully downloaded? If so that likely relates to the moov atom. Briefly, the moov atom contains the index information in the encoded file that the player needs to play the file. Here's a quick summary from that Apple website at bit.ly/moovatom:

> A QuickTime movie file contains information about the movie, stored in a 'moov' atom—which contains one 'trak' atom for each track in the movie, a 'udat' atom for user data, and so on. This information tells QuickTime what's actually in the movie and where it's stored.
>
> QuickTime needs to load the 'moov' atom into the computer's memory in order to play a movie. When you save a self-contained fast-start movie, QuickTime puts the 'moov' atom at the front of the file (it's usually only 1 or 2 Kbytes), followed by the movie data, arranged in chronological order. When you download the file over the Internet, the 'moov' atom arrives right away, so QuickTime can play the movie data as it comes in over the net.
>
> If the 'moov' atom is at the end of the file, QuickTime doesn't know what's in the movie or where it's stored, so it doesn't know what to do with the movie data as it comes in, and the movie can't play until the 'moov' atom arrives at the end of the file.

Though the blog post obviously relates to QuickTime, the moov atom affects Flash playback as well. Here's a blurb from *Kaourantin.net*, the blog maintained by Adobe Developer Tinic Uro (bit.ly/flashh264).

> If you use progressive download instead of FMS make sure that the moov atom (which is the index information in MPEG-4 files) is at the beginning of the file. Otherwise you have to wait until the file is completely downloaded before it is played back. You can use tools like qt-faststart.c written by our own Mike Melanson to fix your files so that the index is at the beginning of the file. Unfortunately our tools (Premiere and After Effects etc.) currently place the index at the end of the file so this tool might become essential for you, at least for now. We are working hard to fix this in our video tools. There is nothing we can do in the Flash Player and iTunes/QuickTime does behave the same way.

Though I haven't experienced these playback delays personally with Silverlight or HTML5, this moov atom location issue likely affects these other formats as well.

Note that you won't encounter this problem when distributing your video via a streaming server because the server can communicate with the player and pass along any necessary playback information. This only occurs when distributing your videos via progressive download because there's no such communication between server and player.

Avoiding the Problem

When producing for QuickTime playback, you can avoid this problem by choosing the Fast Start option in whichever encoding tool you're using. This places the moov atom at the start of the file. Don't choose Fast Start–Compressed Header, because if you later decide to deploy the file via Flash (and likely Silverlight and HTML5), the player may have problems playing the file.

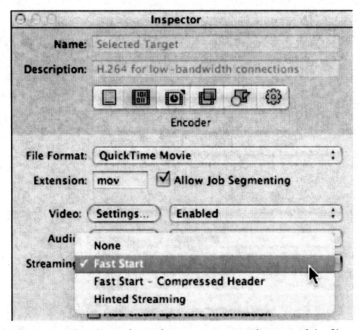

Figure 5-4. Fast Start places the moov atom at the start of the file.

When producing for Flash, most tools don't offer the Fast Start option; they either place the moov atom in the right place, or don't, but either way, you have no control.

Moving the Moov Atom

What to do if you have a file with the moov atom at the back of the file? Well, you can re-encode the file selecting the Fast Start option, or download the QTIndexSwapper from bit.ly/fixmoov. It's a free Adobe Air application that analyzes the file, determines if the header is in the right place and corrects the problem if necessary.

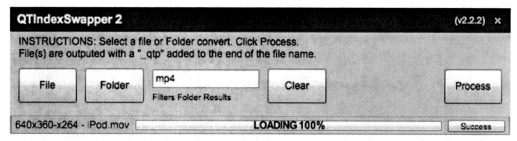

Figure 5-5. Moving the moov atom to the front of the file with QTIndexSwapper 2.

Once you install the application and get it up and running, click File or Folder to add single or multiple files, then click Process. The adjustment only takes a few moments per file, and once you re-upload your files, they should start to play after only a few moments of buffering.

OK, now that we've detailed an issue that can affect all playback environments, let's dig a bit deeper into the specific environments.

Producing for QuickTime Playback

QuickTime Player does best with MOV and MP4 files, so use those container formats and extensions when producing files for the QuickTime Player. I was unable to find specs for what profiles and levels the QuickTime Player will play, though through testing, I learned that the current version (7.6 and higher) will play H.264 video encoded with the Baseline, Main and High profiles. Note that versions prior to 7.2 could not play files encoded with the High profile, but since that version came out in July 2007, and Apple is very aggressive about convincing its users to upgrade, it's likely that the vast majority of potential viewers support the High profile.

Note that since Apple Compressor only supports the Main profile without CABAC, all files actually posted by Apple on its site conform to these specs—there are none using the High profile or CABAC. I don't think it's necessary to emulate this behavior; I just think that Apple is avoiding the embarrassing scenario of being caught using another encoding tool to produce video for its own website.

In my tests, on an 8-core Mac Pro, the QuickTime Player played a 1920x1080p@20 Mbps file encoded using the High profile at Level 5.1 without problem. So long as your file is High profile or lower, it should play in QuickTime. Of course, if you're trying to create a file that will play on all versions of QuickTime Player, even those preceding 7.2, you can use the Main profile, which will result in a file that will also play on the iPad (so long as it's 720p or smaller).

As we learned in the previous section, when you're producing a file for progressive playback, use the Fast Start option, avoiding the compressed header to maximize compatibility with other players. If you're producing for the QuickTime Streaming or Darwin servers, use the Hinted streaming option.

Producing for Flash Playback

This one is simple. The Flash Player supports the Baseline, Main, High, and High 10 H.264 profiles with no levels excluded. Accordingly, when you're producing H.264 video for Flash Player, you can choose the most advanced profile supported by the encoding tool, which is typically the High profile. On the audio side, Flash Player can play AAC Main, AAC Low Complexity, and AAC SBR (spectral band replication), which is otherwise known as High-Efficiency AAC, or HE-AAC (**adobe.ly/h264encode**).

In terms of container format, the Flash Player should be able to play files in the QuickTime, MPEG-4 and 3GP container formats. Here's a blurb from Adobe developer Tinic Uro's blog (**bit.ly/adobeh264**):

> You can load and play .mp4, .m4v, .m4a, and .mov files using the same NetStream API you use to load FLV files now. We did not add any sort of new API in the Flash Player. All your existing video playback front ends will work as they are. As long as they do not look at the file extension that is, though renaming the files to use the .flv file extension might help your component. The Flash Player itself does not care about file extensions, you can feed it .txt files for all it matters. The Flash Player always looks inside the file to determine what type of file it is.

Let's be pedantic and identify the key points he's making one by one.

• As I said, Flash Player should be compatible with QuickTime, MPEG-4 and 3GP files.

• Second, file extension doesn't matter. This is very convenient because neither of the two Flash development environments, Flash Builder or Flash Catalyst, can import MP4, MOV or 3GP files. However, if you change the extension of the same file to .flv (or .f4v), they will load right in. The only exception that I've run into is when you have a compressed header; those files won't load.

File Extension	Mime Type	Description
.f4v	video/mp4	Video for Adobe Flash Player
.f4p	video/mp4	Protected Video for Adobe Flash Player
.f4a	audio/mp4	Audio for Adobe Flash Player
.f4b	audio/mp4	Audio Book for Adobe Flash Player

Figure 5-6. MIME types for Flash-related audio and video files. Image courtesy Adobe.

As mentioned above, F4V has become the container format for Flash video. Exactly what is F4V? It's a new container format based upon the MPEG-4 specification with additional support for metadata and other properties that support playback within the Flash environment. The new container format has four types of files with four different extensions, as shown in Figure 5-6 (bit.ly/wikiflashvideo).

Long story short, so long as you're producing Plain Jane MP4 or MOV files, they should play in Flash Player. If you change the extension from .mp4 or .mov to .f4v, they should still play, unless you've produced with a compressed header. If you're working with an encoder like Adobe Media Encoder that outputs directly in F4V format, output into that format, but recognize that the file may not play in QuickTime Player or Silverlight, even if you change the file extension to .mov or .mp4. If you're trying to create a file that will play everywhere, use .mp4.

Producing for Silverlight Playback

Silverlight should play H.264 files encoded in the .mov, .mp4 and .f4v container formats, and can play back files encoded using the Baseline, Main and High profiles at all levels, and AAC-LC (low-complexity) mono or stereo. To paraphrase Microsoft Silverlight evangelist Ben Waggoner, if QuickTime 7 can play the file, Silverlight can play it.

Microsoft's Expression Encoder software program can encode to H.264 format at up to the High profile, with low-complexity AAC audio, and produces MPEG-4 standard MP4 files during output. This tells me that if you're producing for Silverlight with any other encoding program, the MP4 wrapper is your best option.

Producing for HTML5 Playback

As mentioned, when producing for HTML5, it's best to produce an MP4 file, because MOV or F4V files could trigger playback using either the QuickTime or Flash Player. The H.264 playback capabilities of the three HTML5-compatible browsers—Apple Safari, Google Chrome (for the time being) and Microsoft Internet Explorer 9—is obviously up to the browser vendor; as far as I could find, none of these vendors have published specifications that detail profile and level-type compatibility.

In my tests, all three browsers played a 1080p file encoded using the High profile, with stereo AAC-LC audio. So long as you produce a file that doesn't exceed these parameters, you shouldn't run into any playback problems with any H.264-compatible HTML5 browser.

Changing the File Extension on an H.264 File

I covered this point above when talking about Flash, but let's quickly reiterate. Suppose you produced an H.264 file in the MOV format and want to deploy it in the Flash or even Silverlight environment. If you're hand-coding your player, you can keep the file in the MOV format, and it will play just fine.

On the other hand, tools like Adobe Flash Catalyst or Flash Builder won't import a MOV or MP4 file, only FLV or F4V, so you have to change the file name so the program will recognize the file. There are no absolutes in video production, only tendencies. However, so long as the file that you've produced doesn't have a compressed header, both Flash tools should be able to import the file after you change the extension to .flv or .f4v.

Configuring Your H.264 Streams

Now let's look at the factors to consider when choosing the resolution, frame rate, video data rate and audio encoding parameters. First, we'll see how various configurations of H.264 video play on a variety of computers—no sense creating a stream that looks and plays great on your studio 8-core workstation, but quickly brings a consumer PC to its knees.

Then we'll look at the encoding parameters used by high-profile media and corporate sites, which distribute millions of streams in the course of a year and know which data rates their viewers can successfully stream and the quality those data rates will deliver. Along the way, we'll take a quick look at the frame rate issue and then zero in on the best configuration parameters for the audio component of your video file.

Flash Playback Statistics

Let's start with a quick look at how H.264-formatted Flash files play back on a variety of desktop and mobile Windows and Mac computers. To perform this test, I encoded a single 29.97 fps 720p test file into six different configurations, starting at 320x240 and ending at 720p. Then I tested playback on nine different computers, playing the file for about two minutes and recording CPU utilization figures from Windows Task Manager or Mac Activity Monitor. You can check out the files yourself at bit.ly/flashplayback. I used Mozilla Firefox to play all files on Windows, and Safari on the Mac, checking that each computer had the latest version of both the browser and Flash Player.

To add a bit more detail, I encoded the files so they had a bits-per-pixel value of .08, a metric that you'll learn about later in the chapter. I encoded using Sorenson Squeeze, using my standard H.264 encoding parameters, which included High profile, CABAC enabled, three B-frames, five reference frames, zero slices and quality set to Best.

Windows Tests

I focused my testing on the slower computers that I have in my office or have access to. On my 24-core HP Z800 or 16-core Mac Pro workstations, I can play multiple streams of 1080p video without a hitch. But few, if any, publishers focus on computers like these. So instead, I tested the older workstations in my office—some that I've had since 2003—to get a sense of how the various test configurations would play on these older computers, which make up a much larger percentage of the installed base that producers do care about.

For Windows desktop computers, I tested a computer with the last Pentium 4 processor (the HP xw4100, circa 2003), a computer with the first dual-core CPU (the HP xw4300, circa 2006) and a computer with the first Core 2 Duo processor (the Dell Precision 390, circa 2006). As you'll see in the table, the Precision 390 played video up to 720p without much of a hiccup, so it didn't make sense to test any faster, more modern desktops. All three Windows desktops that I tested were configured with NVIDIA graphics cards that accelerated the playback of H.264-encoded Flash, a nice advantage.

Windows	Notebooks			Desktop		
	2004	2007	2008	2003	2006	2006
File playback	Dell Latitude D800	HP mobile 8710p	Acer Aspire One	HP xw4100	HP xw4300	Dell Precision 390
CPU	1.6 GHz Pentium M	2.2 GHz Core 2 Duo	1.60 GHz Atom	3 GHz Pentium 4	3.4 GHz Pentium D	2.93 GHz Core 2 Duo
GPU	NVIDIA GeForce 4200	NVIDIA Quadro FX 1600M	Intel 945 Express	NVIDIA Quadro4 380XGL	NVIDIA Quadro FX 3450	NVIDIA Quadro FX 3500
320x240 file - CPU %	38	12	37	21	4	2
480x360 file - CPU %	63	14	37	28	9	5
640x360 file - CPU %	75	14	35	36	12	6
640x480 file - CPU %	79	17	57	40	15	8
848x480 file - CPU %	NA	16	NA	49	18	10
720p file file - CPU %	NA	27	NA	78	41	24

Table 5-1. The percentage of CPU required to play back video at these test configurations on these computers.

Notebooks are a different story, primarily because there are so many different configurations, some built for portability, some for power. Here, I tested a Dell Latitude D800 Pentium M-based notebook from 2004, an HP 8710p mobile workstation from 2007 (with a 2.2 GHz Core 2 Duo) and a netbook, specifically the Acer Aspire One with a 1.6 GHz Intel Atom CPU. I present the results in Table 5-1.

Let's start with the notebooks. My rule of thumb is that if more than 50-60% of CPU is required for video playback, other operations become sluggish, and dropped frames or audio stoppages become likely. Numbers in excess of 70% are big red flags signaling likely playback issues. In this regard, the Latitude D800 struggled to play the 640x360 and 640x480 configuration, and shut down completely on files with a larger configuration. However, both the HP 8710p and Acer Aspire One played both of these files without significant problems.

On the desktop side of the equation, all three computers successfully played files up to 848x480 resolution, though the Pentium 4-based xw4100 had problems with the 720p file.

Now let's look at the Mac-based tests.

Mac Tests

I tested three Macs, an oldie-but-still-goodie Dual G5 Power PC, my daughter's 2. GHz Core 2 Duo-based iMac, and a 3.06 GHz Core 2 Duo-based MacBook Pro. Here we see that all three computers sailed through all tests up to and including the 848x480 tests, though the PowerPC started to drop frames at 720p.

Mac			
Launched	2005 (est)	2007	2009
Computer	Dual G5	iMac	MacBook Pro
CPU	Dual 2.7 GHz Power PC	2 GHz Core 2 Duo	3.06 GHz Core 2 Duo
GPU	NVIDIA GeForce 6800 Ultra chipset	ATI Radeon X160	NVIDIA GeForce 9600M
320x240 file - CPU %	23	22	17
480x360 file - CPU %	27	33	17
640x360 file - CPU %	31	30	28
640x480 file - CPU %	37	33	24
848x480 file - CPU %	43	34	39
720p file file - CPU %	73	53	56

Table 5-2. CPU required to play back video at these test configurations.

If you scan through both sets of results, you'll see that all computers except the old Dell note-book played 640x360 video at less than 40% of CPU, while all but the Dell played the 848x480 stream at less than 60% of CPU. In the next section, you'll see how these results compare to the actual file configurations used by the various media and corporate websites that I sampled.

Choosing Resolution and Data Rate

I think it's useful to analyze the video configurations used by large media sites and corpora-tions for several reasons. First, media sites distribute the overwhelming majority of non-UGC content, and many have access to analytics that tell them whether their viewers can access the streams in real time. If eight three-letter networks in the US stream their episodes at a combined audio/video data rate approaching 1 Mbps, this tells us that most broadband con-sumers can successfully stream files encoded at that rate.

In addition, media sites help set quality and size expectations. If your target viewer keeps abreast of news by watching CNN at 640x360, then your 360x180 stream will look miniscule. If you're selling consumer electronic devices, and Apple streams its iPad and iPhone advertisements at 848x480, your 320x240 stream will look positively quaint. Finally, if media sites are producing episodes and other content at 640x480 resolution or higher, this likely confirms that a stream using that configuration will play well on most relevant target computers.

For these reasons, every year or so I visit a few dozen prominent media and corporate sites, download their video files and analyze them. The next few tables show the results of the most recent study, which I'll pull together into a pithy, meaningful recommendation chart after discussing the individual results.

Note that the sites that I sampled don't distribute exclusively in H.264: VP6 is still very widely used and there are one or two Windows Media files thrown in. Since H.264 plays back with less CPU overheard than either other format, and delivers superior quality, the numbers presented in the tables are conservative. That is, the recommended resolutions and data rates should produce a very high quality stream that should play back smoothly on most target platforms.

Let's start with media-related websites.

US Media Sites

These media sites include most three-letter networks (ABC, CBS, NBC, CNN, etc.) plus a smattering of other sites like ESPN and USA Networks.

Networks	Width	Height	Total Pixels	Frame Rate
News (5 of 7 16:9)	577	352	206,757	30
Sports (4 of 4 16:9)	613	345	211,920	27
Full episodes (7 of 8 16:9)	627	367	232,182	28
Excerpts/Previews/Other (3 of 4 16:9)	609	343	210,346	30

Table 5-3. Resolution and frame rate statistics from US media sites.

As you can see, the results are divided into four categories: news, sports, full episodes and excerpts/previews/other. News streams, which are informational, are generally smaller than entertainment-oriented streams. With all streams, 16:9 widescreen is the norm, as is publishing at full frame rates. The most commonly used resolution is 640x360, confirming our findings in the previous section that this configuration will play acceptably well on a very broad range of computers relevant to US media sites.

European Media Sites

These sites are from the UK because most other European media sites are geo-blocked to the US, so I couldn't play and analyze these files.

	Width	Height	Total Pixels	Frame Rate
MyChannelOne	480	268	128,640	25
Financial Times	480	270	129,600	25
BBC News	512	288	147,456	25
Average	**491**	**275**	**135,232**	**25**
BBC Sports	640	360	230,400	25
Sky 1	640	360	230,400	25
Telegraph.co.uk	640	360	230,400	25
Channel 4	720	406	292,320	25
Average	**660**	**372**	**245,880**	**25**

Table 5-4. Resolution and frame rate statistics from across the pond.

From the results we do have, we can see that news streams are again smaller (491x275 average) than sports and general entertainment streams (660x372). All streams were at full PAL frame rate (25 fps), and all are 16:9 widescreen.

US Business to Consumer (B2C) Sites

These results include 17 prominent B2C sites like Nike, Coke, Porsche and Burberry, which I've divided into three classes based on total pixels; under 200K is conservative, 200-400K mid-range, more than 400K aggressive.

	Width	Height	Total Pixels	Frame Rate
Conservative (under 200K)	549	307	169,680	26
Mid-range (200-300K)	658	371	244,777	27
Aggressive (over 300K)	949	497	474,047	25

Table 5-5. Resolution and frame rate statistics from US B2C sites.

Even conservative sites averaged well above 320x240 minimum; overall, resolutions ranged from 480x268 to 1024x572. Again, virtually all streams are full frame rate and 16:9.

US Business to Business (B2B) Sites

On these prominent sites, resolutions ranged from 480x360 to 934x524, but six of seven were 638x432 or larger. All are full frame rate and six of seven were 16:9, indicating a clear bias toward widescreen video.

	Width	Height	Total Pixels	Frame Rate
HP	480	360	172,800	29.97
UPS	640	360	230,400	29.97
FedEx	638	480	306,240	30
Adobe	768	432	331,776	30
Cisco	902	507	457,314	30
GE	934	524	489,416	30
Average	**727**	**444**	**331,324**	**30**

Table 5-6. Resolution and frame rate statistics from US B2B sites.

European Commercial Sites

The next table presents statistics from a number of European B2B sites, including Maersk, Bosch, GlaxoSmithKline and BASF. In this group, conservative was under 200K total pixels, 200-300K mid-range, and over 300K aggressive.

	Width	Height	Total Pixels	Frame Rate
Corp - Conservative (9)	440	273	123,343	24
Corp - Mid-range (4)	692	389	269,599	25
Corp - Aggressive (2)	692	450	309,468	25

Table 5-7. Resolution and frame rate statistics from European B2B sites.

These organizations are generally more conservative than US B2B sites regarding stream resolution, though most produced at 16:9, and all but one at full frame rate.

Resolution Synthesis and Recommendations

OK, let's wrap all this data into a nice pretty bow, complete with recommendations. Specifically, your resolution should match the content of your video; if your content is informational (like news), it's OK to be a bit smaller, but for product or other marketing videos, you should consider larger sizes. In both the US and Europe, the resolution sweet spot for the media is around 640x360. If you produce at smaller sizes, you risk looking wimpy; if you go larger, you'll look bold by comparison.

Organization	Minimum	Recommend	Frame Rate
Media - US News	480x270	640x360	full
Media - US Other	640x360	640x360	full
Media - UK	480x270	640x360	full
B2C - US	480x270	640x360 plus	full
B2B - US	480x270	640x360 plus	full
Euro Corp	320x240	640x360 plus	full

Table 5-8. Resolution and frame rate recommendations.

US corporations should use 480x270 as a minimum, and consider producing at 640x360 or higher to make a big splash. European non-media organizations can be a bit more conservative without sticking out, though in the B2C space, organizations like L'Oreal make a strong impression by producing at 686x386.

What About Mod-16?

Many compressionists recommend using what's called a mod-16 resolution, where both the width and height parameters for each stream are divisible by 16. Why? Because most codecs, including H.264, encode in 16x16 blocks, and if the height and width aren't divisible by 16, the codec will create the full block anyway, adding more pixels to the file, which makes it harder to compress.

For example, a 16x16 video file would require one 16x16 block to encode, while an 18x18 video file would require four: one extra on the right and bottom to encode the extra pixels and one on the bottom right to square out the video stream. Note that all of these extra blocks and pixels are automatically cropped during display, so you never see them anyway, but the pixels must be compressed nonetheless.

Obviously, this is a worst-case scenario, and there are two schools of thought on mod-16. One treats non-mod-16 streams like "ring around the collar," evidencing a total lack of sophistication on the part of the compressionist. However, there are some very relevant arguments against the importance of mod-16. First, not all mod-16 resolutions are a perfect 16:9 aspect ratio, forcing the compressionist to either crop pixels or adapt a non-16:9 aspect ratio, which distorts the video, however slightly. Second, the extra pixels are always at the edges, so the encoder can apply less data to these blocks without a noticeable loss in quality.

In addition, the importance of mod-16 resolutions decreases as the resolution increases because extra 16x16 blocks end up comprising less of the total picture. For example, according to Microsoft's Alex Zambelli, who helped configure the stream resolutions for NBC's Olympic and Sunday Night Football streaming offerings, 320x176 versus 320x180 yields a 9% efficiency advantage, but 1920x1072 versus 1920x1080 yields only a 1.5% improvement.

Finally, 640x360 is the most widely used stream resolution in existence today, and it isn't mod-16. This obviously wouldn't be the case if the lack of mod-16 compliance significantly degraded quality. When theory clashes with reality, go with reality.

Here's what Zambelli had to say regarding his Olympic and Sunday Night Football encodes:

> Most resolutions are mod-16, but in some cases, we had to settle for mod-8 or mod-4 to match a video resolution to a particular video player window size. For example, 720x404 was the Sunday Night Football player video window size, so we matched it with one of the encoded resolutions in order to ensure it played optimally without requiring any scaling.

In summary, when configuring your video, try using a mod-16 configuration, or mod-8 (where both height and width are divisible by 8) if mod-16 won't work. You should never use a resolution that's not at least mod-4.

Choosing the Data Rate

After choosing resolution, the next configuration decision is data rate. The most useful metric for choosing a data rate is the bits-per-pixel of the video file, so let's start there.

Understanding Bits per Pixel

Bits per pixel is the amount of data applied to each pixel in the video file. You compute it by dividing the per-second video data rate (e.g. 500 kbps) by the number of pixels per second (height x width x fps) in the video file. If you're not particularly mathematically inclined, you can use a tool called MediaInfo (shown in Figure 5-7) to do the calculation for you.

Why is bits-per-pixel such a valuable comparative metric? Because it lets you assess the data rate of a file in a comparative way.

	Width	Height	Total Pixels	Frame Rate	Video Data Rate	Bits per pixel
Adobe	768	432	331,776	30	650	0.07
Cisco	902	507	457,314	30	700	0.05
HP	480	360	172,800	29.97	550	0.11
GE	934	524	489,416	30	1200	0.08
UPS	640	360	230,400	29.97	800	0.12
FedEx	638	480	306,240	30	917	0.10

Table 5-9. The bits-per-pixel value of these files lets you compare the compression applied to each video file.

In the table above, though GE configures its video at 1200 kbps and HP at 550 kbps, GE is actually more aggressive, applying less data per pixel (.08) than HP (.11). As you'll see at the end of the chapter, you can also use bits-per-pixel to create general rules to apply to videos across a range of resolutions.

Computing Bits per Pixel

Again, you compute bits-per-pixel by dividing the per-second video data rate (e.g. 500 kbps) by the pixels per second (height x width x fps). Or you can download Mediainfo, a free cross-platform (Windows, Mac, Linux) file analysis tool from mediainfo.sourceforge.net. As you can see in Figure 5-7, MediaInfo also identifies the codec, multiple encoding parameters, data rate, audio configuration and other useful file data. I discuss MediaInfo in more detail in Chapter 13.

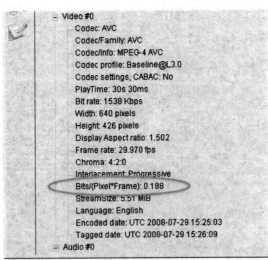

Figure 5-7. MediaInfo is the one tool on every computer in my office.

Using Bits per Pixel

Bandwidth is a major cost for media sites, yet obviously the quality of their video is very important. For this reason, media sites experiment to find a data rate that provides acceptable quality at the lowest possible data rate. By identifying the data rate used by media sites, we can create general rules to apply to media and other sites.

	Total Pixels	Frame Rate	Video Data Rate	Average Bits per Pixel	Bits per Pixel Range
News (5 of 7 16:9)	206,757	30	536	0.092	.076 - .130
Sports (4 of 4 16:9)	211,920	27	803	0.141	.085 - .179
Full episodes (7 of 8 16:9)	232,182	28	866	0.132	.075 - .201
Excerpts/Previews/Other (3 of 4 16:9)	210,346	30	777	0.118	.058 - .174

Table 5-10. The bits-per-pixel value of US media sites.

For example, for relatively low-motion news sites, the average bits per pixel of .092 is significantly lower than higher-motion sports (.141) and other content. For full television episodes, video quality at an average of .132 is sufficient.

If you're looking to gauge a studio's view of the bandwidth enjoyed by its viewers, the average data rate for full episodes is 866 kbps. Add audio at an average of 128 kbps, and you get an overall data rate of just under 1 Mbps. You may not want to configure your video at quite that total data rate, but clearly the three-letter networks think the bulk of their customers can smoothly retrieve and play files encoded at that data rate.

Avoiding Wasted Bandwidth

Table 5-11 presents the bits-per-pixel of prominent B2C brands in the US.

	Total Pixels	Frame Rate	Video Data Rate	Average Bits per Pixel	Bits per Pixel Range
Conservative (under 200K)	169,680	26	721	0.154	.08 - .27
Mid-range (200-300K)	244,777	27	579	0.091	.04 - .14
Aggressive (over 300K)	474,047	25	1,814	0.177	.04 - .44

Table 5-11. The bits-per-pixel value of B2C brand sites.

Mid-range sites show an average bits-per-pixel of .091, which is very close to the average of US news organizations, though the average for aggressive sites is much, much higher. Examining the individual videos in the aggressive group, however, reveals total outliers like Buick at .44 and Apple at .26. These values are way too high, indicating either extraordinarily difficult video to compress or wasted bandwidth. If you drop these two videos from the analysis, the aggressive group average drops to a much more reasonable .089.

One more table and then we'll pull this data together into some general rules.

Bits per Pixel Should Decrease as Resolutions Increase

Table 5-12 presents the bits-per-pixel values of European corporate sites. Notice that the bits-per-pixel values decrease as the resolutions increase. This is because codecs work more efficiently as video resolutions increase, so values that are effective at 640x360 resolution may not be sufficient at 320x180.

	Total Pixels	Frame Rate	Video Data Rate	Average Bits per pixel	Bits per Pixel Range
Corp - Conservative (9)	123,343	24	484	0.163	.095 - .292
Corp - Mid-range (4)	269,599	25	667	0.101	.055 - .139
Corp - Aggressive (2)	309,468	25	712	0.092	.052 - .131

Table 5-12. The bits-per-pixel value of European corporate sites.

Long story short, as video resolutions increase, you can maintain quality at a progressively lower bits-per-pixel value. How much lower? Microsoft evangelist Ben Waggoner has quantified this as the Power of .75 rule, which he defines as follows (in a 2009 email to me):

> Using the old "Power of .75" rule, content that looks good with 500 kbps at 640x360 would need (1280x720)/(640x360)^.75*500=1414 kbps at 1280x720 to achieve roughly the same quality.

Essentially, as the resolution gets higher, a lower bits-per-pixel value will sustain the same quality level. On the off chance that you may not be up to speed on how to compute fractional exponents (I sure wasn't), I'll share tables that lay out this math after a short summary.

Applying Bits per Pixel

Taking all of this data into account, my rule of thumb is that a reasonable bits-per-pixel is around .1 for videos with low to average motion—your basic talking head or news video—ranging to .15 or higher for fast-moving videos. ESPN streams at around .178, which tells me that you might consider going a bit higher for sports videos.

You should apply these values to video configured at 640x360 resolution (16:9) or 640x480 resolution (4:3). According to the Power of .75 rule, for lower-resolution videos, the bits-per-pixel value should increase, while it should decrease for higher-resolution videos. I used these concepts to create Table 5-13, which is computed for videos produced at 30 fps.

30 fps		Low Motion		High Motion	
Width	Height	Data Rate (kbps)	Bits per Pixel	Data Rate (kbps)	Bits per Pixel
16:9					
320	180	244,376	0.14	366,564	0.21
480	270	448,948	0.12	673,421	0.17
640	**360**	**691,200**	**0.10**	**1,036,800**	**0.15**
853	480	1,063,860	0.09	1,595,790	0.13
1280	720	1,955,009	0.07	2,932,513	0.11
1920	1080	3,591,581	0.06	5,387,371	0.09
4:3					
320	240	325,835	0.14	488,752	0.21
400	300	455,368	0.13	683,052	0.19
480	360	598,597	0.12	897,895	0.17
640	**480**	**921,600**	**0.10**	**1,382,400**	**0.15**

Table 5-13. Recommended data rates and bits-per-pixel values for video produced at 30 fps.

The top of the table deals with 16:9 videos, and normalizes the .1 low-motion and .15 high-motion rules at 640x360. I've applied the rule of .75 upward and downward to computer data rates and bits per pixels at other resolutions. The bottom four lines are for 4:3 videos.

Obviously, the data rates and bits-per-pixel values are suggestions. However, if you're producing at much higher than those values, your data rate may be excessive, unnecessarily increasing your bandwidth costs and making your videos harder to smoothly play for those on marginal connections. If you're producing at much lower and your video quality isn't up to par, perhaps you should re-evaluate. Here are tables for 24 fps and 25 fps.

24 fps		Low Motion		High Motion	
Width	Height	Data Rate (kbps)	Bits per Pixel	Data Rate (kbps)	Bits per Pixel
16:9					
320	180	195,501	0.14	293,251	0.21
480	270	359,158	0.12	538,737	0.17
640	**360**	**552,960**	**0.10**	**829,440**	**0.15**
853	480	851,088	0.09	1,276,632	0.13
1280	720	1,564,007	0.07	2,346,011	0.11
1920	1080	2,873,264	0.06	4,309,897	0.09
4:3					
320	240	260,668	0.14	391,002	0.21
400	300	364,294	0.13	546,442	0.19
480	360	478,877	0.12	718,316	0.17
640	**480**	**737,280**	**0.10**	**1,105,920**	**0.15**

Table 5-14. Recommended data rates and bits-per-pixel values for video produced at 24 fps.

25 fps		Low Motion		High Motion	
Width	Height	Data Rate (kbps)	Bits per Pixel	Data Rate (kbps)	Bits per Pixel
16:9					
320	180	203,647	0.14	305,470	0.21
480	270	374,123	0.12	561,184	0.17
640	**360**	**576,000**	**0.10**	**864,000**	**0.15**
853	480	886,550	0.09	1,329,825	0.13
1280	720	1,629,174	0.07	2,443,761	0.11
1920	1080	2,992,984	0.06	4,489,476	0.09
4:3					
320	240	271,529	0.14	407,294	0.21
400	300	379,473	0.13	569,210	0.19
480	360	498,831	0.12	748,246	0.17
640	**480**	**768,000**	**0.10**	**1,152,000**	**0.15**

Table 5-15. Recommended data rates and bits-per-pixel values for video produced at 25 fps.

Conclusion

Now you know how to configure your H.264 video when producing for computers; in the next chapter, we take on producing for Apple iDevices.

Chapter 6: Producing for Apple (and Other) Devices

As the title suggests, the main focus of this chapter is how to produce video for Apple devices, from iPod to iPad 2 with the iPhone and iPod touch in between. At the end of the chapter, I'll briefly cover the Android, BlackBerry, HP webOS (formerly Palm), and Microsoft Window Mobile 7, but the focus is predominantly upon Apple.

Why Apple devices? Several reasons. First, Apple had an early lead in market share, so most web producers prioritize supporting these devices. In addition, Apple has done a wonderful job documenting the requirements of its iDevice family, making it simple for both the writer and the streaming producer.

As an overview, when producing for iDevices, there are two primary use cases: iTunes delivery, and for those devices with wireless capabilities, producing for Wi-Fi or cellular delivery. As you'll see, though the playback specifications of the various devices vary greatly, most producers take a unified approach that analyzes the most efficient way to deliver the highest-possible-quality files to the most relevant group of iDevice users. That's how we'll approach things in this chapter.

Producing for iTunes Delivery

Table 6-1 details the device and playback specifications of all Apple iDevices. The screen resolution of the device is shown on the first line, while the H.264 playback capabilities are detailed on the following lines.

	Original iPod (pre-5g)	iPod Nano/Classic	iPod touch/ iPhone	iPhone 4/iPod touch	IPad 1&2
Device spec					
Screen rez	320x240	320x240	480x320	960x640	1024x768
Aspect ratio	4:3	4:3	16:9-ish	16:9-ish	4:3
Codec spec					
Video codec	H.264	H.264	H.264	H.264	H.264
Max data rate	768 kbps	2.5 Mbps	2.5 Mbps	14 Mbps	14 Mbps
Max video rez	320x240	640x480	640x480	720p	720p
Frame rate	30 fps	30 fps	30 fps	30 fps	30 fps
Profile/level	Baseline to Level 1.3	Baseline to Level 3.0	Baseline to Level 3.0	Main to Level 3.1	Main to Level 3.1
Audio codec	AAC-LC	AAC-LC	AAC-LC	AAC-LC	AAC-LC
Max data rate	160 kbps	160 kbps	160 kbps	320 kbps	320 kbps
Audio params	48 kHz, stereo	48 kHz, stereo	48 kHz, stereo	48 kHz, stereo	48 kHz, stereo
Container formats	m4v/mp4/mov	m4v/mp4/mov	m4v/mp4/mov	m4v/mp4/mov	m4v/mp4/mov

Table 6-1. Device specifications for iDevices.

In most instances, iTunes won't load files that exceed a device's playback capabilities, so if you try to download a 640x480 file to a first-generation iPod, iTunes will present an error message that looks something like Figure 6-1.

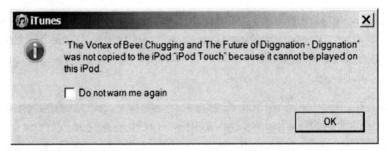

Figure 6-1. This is what happens when you try to synch a file that exceeds the iDevice's playback capabilities.

Scanning Table 6-1, you see that there are multiple levels of playback capabilities: 320x240 for pre-5G iPods, 640x480 for later iPods and the iPhone and iPod touch devices (pre-4G), and 720p for iPhone 4/iPod touch 4G and the iPad 1&2 (which I'll just call the iPad from here on out). Producing at 320x240 will deliver 100% compatibility but a poor-quality experience on newer devices, particularly the iPhone/iPod touch 4G and iPad.

This raises the following three questions:

- Should you abandon older iPods?

- Should you try to support all post-iPod 5G devices with a single stream?

- Should you distribute multiple files?

To answer these questions, I downloaded 48 podcasts from iTunes from 34 different producers, downloading the most popular podcasts as designated by iTunes, several from three-letter networks and many from popular technology sites like CNET.

Once I found an episode from a specific producer, I searched iTunes to determine if other episodes, at other resolutions, were available. For the purposes of this survey, all downloads were free; I'll discuss paid TV show episodes and music videos below. Let's address our three questions in order.

Should You Abandon Older iPods?

Here's what I found:

- Nine of 48 files were 320x240 or smaller. This meant that nine out of the total 34 producers (or 26%) continued to serve the pre-5G market with 320x240 H.264-encoded files. These included CNN, CNBC, Oprah, Tekzilla (from Revision 3), CBS and videos from President Obama.

- One producer also released a 640x480 file using the older and lower-quality MPEG-4 codec, which should play on pre-5G iPods. So 29% of overall producers still addressed the oldest iPods.

- When I performed a similar survey in 2008, 56% of producers produced at 320x240 or smaller.

- Only three of these producers released the same show at higher resolutions. Emphasis is on the same show because two producers, CNBC and CBS, produced other shows at higher resolutions, but not the same show. For six of nine, this low-resolution file was the only file that they made available to all iDevices.

My conclusion? That producers seeking the broadest possible audience still address the pre-5G market.

Should You Support Post-5G Devices with a Single Stream?

Eighteen of the 48 files were 640x480 or 640x360, the largest single group. However:

- Nine of the 18 also produced at a higher or lower resolution, so for nine of 34 producers (26%), this file was the sole offering.

- Of the nine that also offered an alternative file, one produced at 320x240 or smaller, addressing the pre-5G market. Six produced at 720p and one at 480x270, which is customized for pre-4G iPhone/iPod touch devices. One producer offered both 720p and 320x240 files.

My conclusion? A single file at one of these resolutions should be fine for casual producers; those seeking maximum impact on newer devices should consider multiple streams.

Should You Distribute Multiple Files?

• Twelve of 34 producers (35%) delivered the same show at multiple resolutions. This included Oprah, CNET and Tekzilla (Revision 3), but didn't include ABC, which produced solely at 640x360, CNN (320x240), or MSNBC and NBC (424x240).

• Four producers provided different shows at different sizes. These included CBS, CNBC, CNET and Revision 3. For CBS, that meant that *Katie Couric's Notebook* (320x240) was different from the *Evening News* (384x216), while for CNBC, it meant that Cramer (640x480) got a bigger screen than typical news-related announcements (320x240).

• In terms of presentation, some producers uploaded different resolutions to the same channel (Oprah) while others created different channels for the different resolutions (Revision 3).

Conclusions? Since delivering multiple files costs you nothing but encoding time and storage space, give it a shot. There are also no style points for consistency, so don't feel like you need a consistent policy for all content distributed via iTunes.

Choosing the Optimal Resolution

When choosing whether to produce at 320x240 or higher, consider the following:

• **Content.** Simple talking head content doesn't really benefit from the larger resolution. Entertainment-oriented content or screencam-based content will.

• **Target market.** If you're trying to reach children or young adults, it's unlikely that many have older iPods that don't play files larger than 320x240. If you're targeting mature adults, the penetration of these older devices may be higher.

• **Serving the iPad.** A file produced at 640x360 will look acceptably good on the iPad; files produced at 320x240 won't.

• **Differentiate premium content.** If you have premium shows in your stable, consider differentiating them by resolution.

• **Avoid resolutions that won't play on lower-resolution devices but aren't optimal on higher-resolution devices.** 480x270 made sense before the iPad and iPhone/iPod touch 4G because 480x320 was the largest available iDevice display resolution. Now, 640x360 is a better choice because it will look better on high-resolution iDevices and doesn't exclude any devices that would play 480x270. Another oddball formats was 850x480, which wouldn't play on most low-resolution iDevices, but wouldn't look optimal on the iPad or iPhone/iPod touch 4G. Twelve of 13 producers that produced at higher than 640x480 used 720p, which looks good on compatible iDevices and on your computer.

Working With 16:9 Footage on iPods with a 4:3 Display

iPods with 4:3 displays default to a center-cut display that cuts off the sides of 16:9 videos. For example, in the video shown below, the grayed edges on the right and left sides of the video won't be visible in an iPod with a 4:3 display in the default configuration.

Figure 6-2. 16:9 video displayed on a 4:3 iPod display in the default configuration.

When producing 16:9 videos for display on 4:3 iPods, you have multiple options:

- **Shoot center-cut.** When working with real-world video, you can frame your camera so that the center-cut region contains the critical content. This is obviously what Oprah does in Figure 6-3, showing a frame grabbed from one of her iTunes podcasts.

Figure 6-3. This video is framed for 4:3 center-cut display, so if the edges are cut off, no critical content is lost.

- **Letterbox your video.** CBS produces 16:9 video in a 640x480 4:3 window by letterboxing on top and bottom. This works well for 4:3 displays, which will display the entire video rather than cutting off the sides, though the video will look tiny compared with the center-cut approach. In addition, 16:9 displays like the iPhone/iPod touch devices may shrink the video window as compared with native 16:9 video. Be sure to test how the video plays on both 4:3 and 16:9 iDevice displays before adopting this approach.

Figure 6-4. CNBC displays its 16:9 video within a 4:3 window with bars on top and bottom.

- **Educate the viewer.** Alternatively, you can advise viewers watching on 4:3 iPods to disable the Fit to Screen option and show the entire video, with black bars on the top and bottom of the video. Disable this option by clicking Videos > Settings > Fit to Screen > Off.

Recommended Encoding Parameters

Essentially, when producing for iTunes, you should choose one of three file sizes: 320x240 for ultimate compatibility; 640x360 for compatibility with low-resolution iDevices, excluding the pre-5G units; and 720p for the iPad and iPhone/iPod touch 4G. To determine the optimal encoding parameters, I compared the parameters used in Apple Compressor presets with those used in files that I downloaded. In two of three instances, the presets came close to matching the actual results; in the other, the two were very far apart.

Encoding 320x240 Podcasts

Table 6-2 shows how Compressor's encoding parameters for 320x240 video compared with those used by the files in the survey I conducted.

	320x240 - Compressor	Survey
Video codec	H.264 codec, Baseline profile	H.264 codec, Baseline profile
Data rate average/max	600/768 kbps	528 kbps (average)
Key frames	120	120
Frame rate	match source	match source
Audio	AAC Low	AAC Low
Data rate	128 kbps/stereo	111 kbps/stereo
Extension	.m4v	.m4v

Table 6-2. Producing for 320x240 resolution; Compressor compared with iTunes files analyzed in the survey

Compressor's parameters are close to those used in the survey, indicating that Apple's view of reality matched reality. Accordingly:

- If you're producing in Compressor, use the template as is.

- If you're producing in another encoding tool, make sure that the encoder's low-resolution iPod template reasonably conforms to the Compressor settings. If there is no low-resolution iPod preset, build one using the data in Table 6-2 as a guide.

Encoding 640x360 Podcasts

Table 6-3 shows how Compressor's encoding parameters for 640x360 video compared with those used by survey participants.

	640x480 - Compressor	Survey
Video codec	H.264 codec, Baseline profile	H.264 codec, Baseline profile
Data rate average/max	1.5/2.5 Mbps	1.319 Mbps
Key frames	120	120
Frame rate	match source	match source
Audio	AAC Low	AAC Low
Data rate	128 kbps/stereo	114 kbps/stereo
Extension	.m4v	.m4v

Table 6-3. Producing for 640x360 resolution; Compressor compared with iTunes files analyzed in the survey.

Compressor's parameters are again relatively consistent with those used in the survey. Accordingly:

• If you're producing in Compressor, use the template as is.

• If you're producing in another encoding tool, make sure the encoders' 640x360 iPod template reasonably conforms to the Compressor settings. If there is no such preset, build one using the data in Table 6-3 as a guide.

Encoding 720p Podcasts

Table 6-4 shows how Compressor's encoding parameters for 720p video compared with those used by survey participants.

	720p - Compressor	Survey
Video codec	H.264 codec, Baseline profile	H.264 codec, 11 of 12 are Main or High profile
Data rate average/max	10/14 Mbps	2.845 Mbps
Key frames	120	120
Frame rate	match source	match source
Audio	AAC Low	AAC Low
Data rate	256 kbps/stereo	134 kbps/stereo
Extension	.m4v	.m4v

Table 6-4. Producing for 720p resolution; Compressor compared with iTunes files analyzed in survey.

Compressor's parameters are very inconsistent with those used in the survey. Accordingly:

• Though their specs specify the Main profile, the iPad/iPhone/iPod touch 4G have played files encoded with the High profile in my tests. Still, I recommend the Main profile since the difference in quality between the Main and High profile is much smaller than Baseline to Main, and because there may be High profile configurations that don't play on those devices.

• Apple's suggested data rate is very, very high, clearly more than necessary. Consider scaling that back to around 3-4 Mbps.

• Apple's suggested audio data rate is too high for most content that isn't a music video. Use 128 or 160 kbps maximum.

• If you're producing in Compressor, create a custom template based upon the parameters shown on the right. If you're producing in another encoding tool, make sure the high-resolution iPod template conforms to the settings shown on the right, using the Main profile, or create your own using the same data.

As mentioned above, all the podcasts detailed in the previous charts were available for free. How do the rules change when producers charge for their videos? I look at two cases below, music videos and television episodes.

Encoding Music Videos

I downloaded six music videos to determine the parameters used in these instances, and I found a great degree of uniformity as shown in Table 6-5.

Music Videos	Width	Height	Total Pixels	Data Rate (kbps)	Frame Rate	Audio Data Rate (kbps)	Channels	Bits per Pixel
Eminem/Rihanna	640	256	163,840	1,457	23.976	256	stereo	0.371
Ke$ha	640	360	230,400	1,565	23.976	251	stereo	0.283
Lady Gaga	640	464	296,960	1,514	23.976	249	stereo	0.213
Michael Jackson	640	464	296,960	1,512	29.97	256	stereo	0.170
Shakira	640	352	225,280	1,533	25	251	stereo	0.272
Taylor Swift	640	352	225,280	1,513	23.976	248	stereo	0.280

Table 6-5. Encoding parameters for music videos.

The key points here were:

- All videos were 640x480 or 640x360-ish. Producers ignored the oldest iPods but didn't customize for the higher-resolution devices, at least not yet.

- The target audio for all songs was 256 kbps stereo, though sometimes the effective rate was lower. I doubt listeners could discern between 128 and 256 kbps on their cute ear-buds, but clearly the music industry believes that 256 kbps delivers superior quality.

Encoding TV Episodes

I downloaded five HD shows and several SD shows, and I was surprised to learn that iTunes downloaded an SD show for four of the five HD shows, without any separate action on my part. The SD shows just appeared in the same folder as the HD show.

Figure 6-5. Four of the five HD shows that I downloaded also provided an SD show like the CW's Nikita.

There was a great deal of uniformity among HD shows, as you can see in Table 6-6.

HD TV Episodes	Width	Height	Total Pixels	Data Rate (kbps)	Frame Rate	Audio Data Rate (kbps)	Channels	Bits per Pixel
House	1280	720	921,600	3,871	23.976	160	stereo	0.175
Lie to Me	1280	720	921,600	4,298	23.976	157	stereo	0.195
Nikita	1280	720	921,600	4,141	23.976	157	stereo	0.187
Back to School	1280	720	921,600	4,273	23.976	156	stereo	0.193
Law and Order	1280	720	921,600	4,116	23.976	155	stereo	0.186

Table 6-6. Configuration options used with HD TV episodes.

Specifically:

- All files were 720p.

- All were 23.976.

- All were produced between 3.9 Mbps (.175 bits-per-pixel) and 4.3 Mbps (.195 bits-per-pixel).

- Audio was all produced at 160 kbps stereo, with the effective bit rate slightly lower in most instances.

Again, given these values, if the bits-per-pixel value of your videos exceeds .2, you're almost certainly encoding at an unnecessarily high data rate.

What about SD shows? I knew you'd ask, so I included Table 6-7.

SD TV Episodes	Width	Height	Total Pixels	Data Rate (kbps)	Frame Rate	Audio Data Rate (kbps)	Channels	Bits per Pixel
House	640	480	307,200	1,235	23.976	123	stereo	0.168
Law and Order	640	480	307,200	1,459	23.976	124	stereo	0.198
Bad Girls Club	640	480	307,200	1,460	23.976	128	stereo	0.198
Nikita	640	480	307,200	1,520	23.976	125	stereo	0.206
Back to School	640	480	307,200	1,518	23.976	125	stereo	0.206
Real Housewives of B Hills	640	360	230,400	1,539	29.97	122	stereo	0.223
Real Housewives of DC	640	480	307,200	1,424	29.97	106	stereo	0.155

Table 6-7. Configuration options used with SD TV episodes.

Again, there's a great deal of uniformity, including:

- All shows are encoded at either 640x480 or 640x360 to play on all post-5G iPods.

- Most shows were produced at 23.976 fps, with Real Housewives produced at 29.97.

- All shows were encoded at between .155 bits-per-pixel and .223 bits-per-pixel.

Interestingly, those shows rendered at 640x480 were produced at an anamorphic aspect ratio, so they displayed at 16:9 on computers and devices. This is shown in Figure 6-6, from *House*, which was encoded at 640x480 (see Format), but displayed at 853x480 (see Normal Size), and loaded fine on my two-year-old iPod nano. This is a neat trick that maintained compatibility on iPods capable of playing 640x480 video, while producing a larger playback experience on computers or larger-screen iDevices.

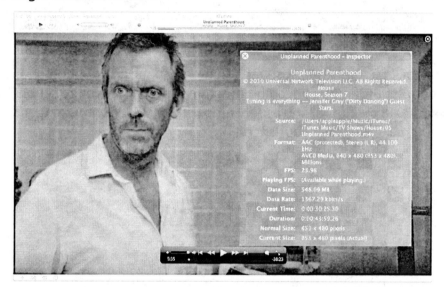

Figure 6-6. House *was encoded at 640x480 resolution, but displayed in 16:9.*

To explain, the other alternative is to produce the file at 640x360, which was the course taken by the producer of *Real Housewives of Beverly Hills*. Obviously, both files will play on 640x480-capable iDevices, but when played on a computer, the 640x360 file will play at that resolution, while the anamorphic 640x480 file would display at the much larger 853x480. Messing around in Compressor, I was able to create the same result with my standard HD test file, as shown in Figure 6-7.

Figure 6-7. Duplicating the same effect in Compressor.

To achieve this look in Compressor, I used the Geometry pane adjustments shown in Figure 6-8. Though controls vary among encoding programs, most allow you to set both a target resolution (640x480 in the figure) and aspect ratio (HD 960x720). Through trial and error, you should be able to make it work in your encoding tool.

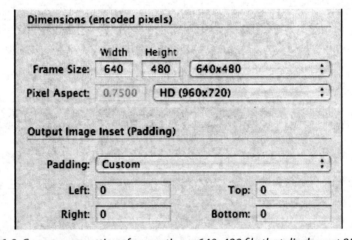

Figure 6-8. Compressor settings for creating a 640x480 file that displays at 853x480.

That wraps up producing for iTunes distribution. Fortunately, producing for streaming to iDevices is a much simpler section.

Producing for Wi-Fi or Cellular Delivery

The optimal technique for delivering to iPhones and iPod touch devices via Wi-Fi or cellular delivery is via Apple's HTTP Live Streaming, which I cover in Chapter 7. This technique allows you to stream adaptively to iPads, iPhones and iPod touch devices, varying stream quality to match the playback capabilities and connection speed enjoyed by the device.

If you're not using HTTP Live Streaming, you have multiple options, including:

- Serving iPad visitors within the browser window via HTML5. In this case, you likely will want to serve a relatively high-quality file to these visitors, and a lower-quality stream to iDevice visitors served via a mobile site.

- Directing all iDevice visitors—including those with iPads—to a mobile site, in which case you likely will serve one lower-quality file to all iDevices.

There are no right or wrong configurations, but consider those shown Table 6-8, which are derived from Apple's Technical Note on HTTP Live Streaming to iDevices (bit.ly/HTTPLive).

	iPad-Only Stream	iDevice Mid-Quality Streaming	iDevice Low-Quality Streaming
Video			
Resolution	640x360	400x224	400x224
Frame rate	29.97	29.97	1/3 (e.g. 10 fps with 29.97 footage
Profile	Baseline/3.1	Baseline/3	Baseline/3
Bit rate control	CBR	CBR	CBR
Video data rate	600 kbps	400 kbps	110 kbps
Key frame interval	3 seconds	3 seconds	3 seconds
Audio	AAC-LC, 40 kbps, mono, CBR	AAC-LC, 40 kbps, mono, CBR	AAC-LC, 40 kbps, mono, CBR

Table 6-8. Recommended encoding parameters for Apple iDevices connecting via Wi-Fi or cellular.

- For iPad-only streams, consider the eponymous column on the left. This roughly corresponds with the lowest data rate 640x360 file recommended by Apple in the Tech Note, which should look very good on the iPad and should play smoothly on many, if not most, iPads.

- When serving iDevices with a single file from a mobile site, the Mid-quality stream conforms to Apple's highest-quality 400x224 stream in the Tech Note, which will look very good on iPhone/iPod touch devices, with passable quality on the iPad. At a combined data rate of 440 kbps, however, some cellular connections may drop frames and pause during playback.

- If smooth playback is your number one goal, consider the Low-quality stream, which will look awful on most devices, but should stream smoothly on all but the flakiest of connections.

Again, nothing magical about these numbers, but they appear to be logical starting points that you can adjust after trying them out for a while.

Producing for Other Mobile Platforms

As mentioned at the top, Apple has done the best job documenting how to produce for its iOS platform, but the other vendors are starting to step up as well. In these next sections, I'll identify the resources available to help those attempting to deliver to other platforms, and briefly review the codecs and configurations supported by the various platforms.

When you look at producing for a platform, you care about at least three aspects: codec, container format and delivery protocols. Not all sites provide information about all three aspects, but I'll summarize what was there.

Android

Android is a "software stack" for mobile devices that includes an operating system and other components that are produced and maintained by Google. If you click over to bit.ly/androidvideospecs, you'll see the formats supported by the Android operating system itself, with hardware or software developers free to add support for additional codecs, or enhanced support for supported codecs.

In terms of codecs, Android 3.0 supports H.263, the Baseline profile of H.264 (starting with Android 3.0), and the Simple Profile of MPEG-4, with support for Low Complexity AAC, AAC+ and enhanced AAC+ audio up to 160 kbps. All three of these standards-based codecs can be delivered in either 3GPP (3GP) and MPEG-4 (MP4) container formats, but there is no support for raw AAC audio files. If you're delivering to the Android, don't wrap your video in QuickTime (MOV) or Flash (F4V).

Starting with version 2.3.3, Android also supports the VP8 video codec and the Ogg Vorbis audio codec in a WebM container format for audio and video, or an Ogg container format for audio only.

Google gave two sets of encoding recommendations on the Android site, as presented in Table 6-9. The low-quality video recommendation definitely lives up to its name, and would produce dismal quality, even though the bits-per-pixel computes to a relatively high .184.

	Low-Quality Video	High-Quality Video
Video codec	H.264 Baseline profile	H.264 Baseline profile
Video resolution	176x144	480x360
Video frame rate	12 fps	30 fps
Video bit rate	56 kbps	500 kbps
Audio codec	AAC-LC	AAC-LC
Audio channels	mono	stereo
Audio bit rate	24 kbps	128 kbps

Table 6-9. Encoding recommendations for Android devices.

In contrast, the high-quality video should actually look pretty good, with a decent resolution and bits-per-pixel of just below .1. Again, remember that these are lowest-common-denominator, software-only numbers. I'm sure that many Android devices will be capable of playing back much higher-quality, but these should be the floor.

In terms of network protocols, Android supports RTSP, HTTP progressive streaming and, starting with Android 3.0, Apple's HTTP Live Streaming. I'm not aware of anyone who has actually delivered adaptive video to the Android platform using HTTP Live Streaming, but I'm sure it's coming. Good move by Google; no sense reinventing that particular wheel.

BlackBerry

BlackBerry does a great job documenting the individual capabilities of all of its phones; unfortunately, there are a lot of phones, so it's tough to find a one-size-fits-all solution. You can download a list of supported formats by phone at **bit.ly/blackberryvidspecs.** I counted 34 phones in nine models.

Scanning through the specs for each phone, it appears that all would play video encoded with the MPEG-4 codec, Simple profile, at 320x240 resolution, 24 fps and a maximum data rate of 768 kbps. The least powerful phone also supported Low Complexity AAC audio, as well as AAC+ and eAAC+, though no audio data rate maximum was stated. If you're working with video shot at 24 fps, this recommendation is a natural, but if your video is 30 fps, I'd avoid 24 fps, which could look choppy, because one out of every six frames is dropped. Instead, I'd encode at 15 fps.

You can encode files into the MPEG-4 (.mp4, m4a,) 3GP (.3gp) and QuickTime (.mov) container formats, and deliver via the Real Time Streaming Protocol (RTSP). You can also wrap the MPEG-4 video in the AVI container format, and deliver Windows Media Video to the lowest-common-denominator BlackBerries, but since few other devices accept that format, I'd recommend sticking with the MPEG-4 container.

At the other end of the spectrum, the most powerful BlackBerry phones can play video encoded into H.264 format, Baseline profile, with a maximum resolution of 480x360, 30 fps, at a maximum data rate of 2 Mbps. Supported container formats for this higher-end video were MPEG-4, 3GP and 3GP2.

HP webOS

The HP webOS is the next-generation operating system for devices formerly known as Palm devices, plus some new ones like the cool-looking HP TouchPad (bit.ly/hptouchpadspecs). You can find the video playback specs for webOS devices at bit.ly/palmvideospecs.

As an overview, you can play video in the webOS either within the webOS Video Player or via HTML5. Either way, webOS can play the codecs identified in Table 6-10 at the specified parameters. HP doesn't mention maximum audio parameters, but it recommends encoding disk-based files using AAC-LC audio at 160 kbps, which I'll take to be the maximum.

	H.264	MPEG-4
Profile	Baseline	Simple
Level	3	5
Video resolution	640x480	640x480
Video frame rate	30 fps	30 fps
Video bit rate	1.5 Mbps	1.5 Mbps

Table 6-10. Encoding maximums for HP webOS devices.

For local playback, HP recommends the following configurations:

- **4:3 content.** H.264 at 480x360@30 fps, with video at 1.5 Mbps and AAC 44 KHz stereo audio encoded at 160 kbps.

- **16:9 content.** H.264 at 480x270@30 fps, with video at 1.5 Mbps and AAC 44 KHz stereo audio encoded at 160 kbps.

Both the audio and video recommendations feel high to me, with the 4:3 video producing a bits-per-pixel value of a hefty .289, with the 16:9 recommendation an even more portly .385. I'd recommend starting at about 520 kbps for the 4:3 video and 400 kbps for the 16:9 video, which would put your bits-per-pixel values around the .1 magic number. For talking head audio, I'd try 64 kbps mono; for high-quality stereo music, I'd max out at 128 kbps. According to the HP documentation, you can use the MPEG-4 (MP4, M4A, M4V), QuickTime (MOV) and 3GP and 3G2 container formats for the audio/video files.

The HP webOS supports both HTTP Progressive Download and the RTSP protocol, with Table 6-11 containing HP's configuration recommendations for each protocol.

	H.264	MPEG-4	MPEG-4
Profile	Baseline/Main/High	Simple	Advanced Simple
Resolution/frame rate	720x480 @ 30 fps 720x576 @ 25 fps	800x600 (720p for HD capture devices) @ 30 fps	800x600 @ 30 fps
Average data rate	2 Mbsp	2 Mbsp	2 Mbsp
Peak data rate	27 Mbps	27 Mbps	27 Mbps
Bit rate control	CBR/VBR	CBR/VBR	CBR/VBR
Audio codec	AAC-LC	AAC-LC	AAC-LC
Audio channels/samples	stereo/48 kHz	stereo/48 kHz	stereo/48 kHz
Audio bit rate	320 kbps	320 kbps	320 kbps
Container formats	MP4, MV, 3GP, 3G2	MP4, MV, 3GP, 3G2	MP4, M4V

Table 6-11. Encoding recommendations for progressive and streaming delivery to HP webOS devices.

While the high-bandwidth progressive download recommendation looks kosher, the two low-bandwidth recommendations look exceptionally conservative. If you decide to support this platform (or should I say when you decide to support this platform), I'd buy a device or two and run multiple tests before adapting these recommendations as is.

Microsoft Windows Phone 7

Windows Phone 7 is the latest mobile operating system from Microsoft that was launched in the United States and Europe in late 2010. Microsoft's minimum device requirements, as published on *Wikipedia* (en.wikipedia.org/wiki/Windows_Phone_7), include at least a 1 GHz ARM or better CPU, a graphics chip capable of GPU acceleration and 256 MB of RAM, which makes for a pretty potent media player.

Table 6-12 details Windows Phone 7's playback capabilities regarding the MPEG-4 and H.264 codecs, with data pulled from a Microsoft document titled, "Supported Media Codecs for the Windows Phone" (bit.ly/windowsphonevidspecs).

Windows Phone 7 supports a wide range of codecs not shown in the table, including Windows Media Video, though encoding in MPEG-4/H.264 produces a file that's reusable in more scenarios. Microsoft doesn't identify the supported protocols, but it does mention the supported delivery scenarios for each codec. At a minimum, Windows Phone 7 can play files encoded with the listed codecs when attached to email and MMS messages, when delivered or played via Microsoft's MediaElement API, when playing in the device's Media Player, and when streaming over the Internet in Internet Explorer Mobile. Check the document for additional details.

	H.264	MPEG-4	MPEG-4
Profile	Baseline/Main/High	Simple	Advanced Simple
Resolution/frame rate	720x480 @ 30 fps 720x576 @ 25 fps	800x600 (720p for HD capture devices) @ 30 fps	800x600 @ 30 fps
Average data rate	2 Mbsp	2 Mbsp	2 Mbsp
Peak data rate	27 Mbps	27 Mbps	27 Mbps
Bit rate control	CBR/VBR	CBR/VBR	CBR/VBR
Audio codec	AAC-LC	AAC-LC	AAC-LC
Audio channels/samples	stereo/48 kHz	stereo/48 kHz	stereo/48 kHz
Audio bit rate	320 kbps	320 kbps	320 kbps
Container formats	.mp4, .m4v, .3gp, .3g2	.mp4, .m4v, .3gp, .3g2	.mp4, .m4v

Table 6-12. Supported MPEG-4/H.264 streams in Windows Phone 7.

In a March 24, 2011, *StreamingMedia* webcast, Alex Zambelli, a Microsoft Media Technology evangelist, added that Windows Phone 7 also supports Smooth Streaming (covered in Chapter 7) using the H.264/AAC-LC or HE combination, but that all streams must have the same resolution.

Conclusion

That's iDevices via iTunes and single-file streaming to a range of mobile devices; next up is adaptive streaming.

Chapter 7: Adaptive Streaming

Adaptive streaming technologies make multiple video streams available to your viewers and dynamically switch streams to adapt to changing connection speeds, CPU use statistics and other heuristics. Adaptive streaming produces the best possible experience for viewers watching on powerful playback stations over high-bandwidth connections while delivering a lower-quality, yet still optimum experience for viewers on low-power devices with spotty connections. If streaming video is mission-critical to your organization, you need to be thinking about implementing some form of adaptive streaming.

Between the chapter title and the compelling lead paragraph, it's probably no surprise that this chapter covers adaptive streaming. I'll start with a brief overview of how adaptive streaming works, and then I'll identify the available technologies and review some concepts to consider when choosing between them. Then, for the major technologies, I'll review implementation details, including where to find published recommendations from each vendor, and I'll provide an overview of how to set up and use each technology.

Next, I'll present a low-level look at how actual producers are using the respective technologies, which will fill in gaps like whether you should produce with VBR or CBR, what key frame interval should you use, and how many streams you should actually produce. I'll conclude with a brief look at Scalable Video Coding, which hasn't been commercially implemented yet, but is a technology that you definitely need to know about.

Technology Overview

Figure 7-1 is a diagram from an Inlet Technologies white paper titled "Powering Smooth Streaming with Inlet Technologies" (bit.ly/inletwhitepapers) that shows the multiple components of an adaptive streaming system—in this case, a system using Inlet's Spinnaker encoder to produce the streams from a live event feed and delivering via Microsoft's Smooth

Streaming technology. Though the diagram details a live event, you can use all adaptive streaming technologies to deliver on-demand video files as well.

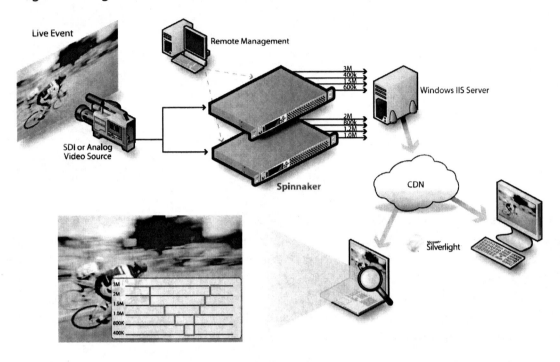

Figure 7-1. Using Inlet Technologies' Spinnaker encoder to produce multiple streams for a Microsoft Smooth Streaming event. Usage courtesy Inlet Technologies.

The components are clear in the picture. You encode multiple streams using either a hardware or software encoder, and then upload the streams to a server, which can be a streaming server as shown in the Figure 7-1, but can also be just a Plain Jane HTTP web server. Then, the server makes the streams available to the various players over the open IP network and adapts the stream to changing conditions during playback. That's what you see on the bottom left of Figure 7-1, with the data rate graph stepping the viewer through the various streams.

How does the system know when to change streams? This depends upon multiple factors, which can include:

- **Buffer conditions.** Each player has a playback buffer—say 5 seconds of video data—that ensures that if the connection stops for intermittent periods, playback will continue. An adaptive streaming technology can monitor the buffer, switching to higher-bandwidth streams when the buffer stays full for a certain duration (indicating a robust connection) or drop to a lower-bandwidth streams when the buffer drops under certain levels (say 2 seconds).

- **Client bandwidth.** For technologies that monitor effective client bandwidth, the system can adjust the stream to match the effective bandwidth.

- **CPU use.** Systems can monitor dropped frames, which are an indication of the CPU power available for video decode. For example, even if the buffer was full and bandwidth sufficient for a higher-quality stream, the system wouldn't switch if the player was currently dropping frames, since a higher-bandwidth, higher-quality stream requires more CPU to decode, and would cause more dropped frames.

- **Playback window size.** Some systems monitor playback size to determine whether to switch to a higher-quality stream. For example, if the video is being played back in a 640x360 window—even if buffer, bandwidth and CPU status indicate the ability to retrieve and playback a higher-bandwidth stream—the system wouldn't switch to a 720p stream because the viewer wouldn't notice the quality difference within the 640x360 viewing window, essentially wasting the additional bandwidth. If the viewer switched to full screen, however, the system would then switch to the higher-quality stream.

Where does the switching logic reside? Typically, on the player since that's the only program actually running on the viewing station.

What happens when the player determines that it's time to switch streams? This depends upon the type of system. In a server-driven system like Adobe's RTMP-based Dynamic Streaming, the player requests a different stream from the server, and the server is in charge of delivering the higher- or lower-quality stream.

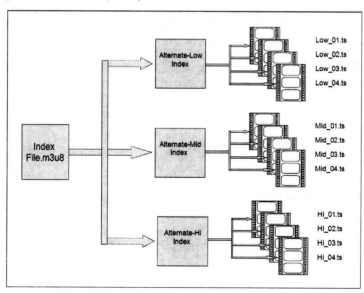

Figure 7-2. The manifest file identifies the different files and their locations. From an article that I wrote for Millimeter *Magazine, and usage is courtesy* Millimeter.

In HTTP-based systems like Apple's HTTP Live Streaming (which also does on-demand), the system depends upon multiple index files that reside on an HTTP server. With HTTP Live Streaming, the index file has an .m3u8 extension, as shown in Figure 7-2. The player is pointed to the index file when the viewer clicks on the web page link to play the video.

All HTTP-based adaptive technologies split their files into 2 to 10-second chunks, which are the video chunks on the right in the Figure 7-2 (Hi_01.ts, Hi_02.ts, etc.). The manifest file identifies the available files and their locations on the HTTP server and refers to specific chunks of each stream. The player might start out playing the lowest-quality stream (Low_01.ts), but if it determines that it can handle a higher-quality stream, it would check the manifest file for the next chunk of a higher-quality stream (say, Mid_02.ts) and retrieve and play that chunk. In an on-demand presentation, the manifest file is static, while in a live event, the manifest file is continually updated with the identity and location of all streams and chunks of streams.

For the following discussions of technology alternatives, it's useful to identify the common components of all adaptive streaming technologies.

- **Encoder.** The encoder can be live or on-demand and can be proprietary to the adaptive streaming system or third party, but some tool must encode the files to be distributed.

- **Server.** The server can be a standard HTTP web server or proprietary RTMP or RTSP streaming server, but the video files must reside somewhere (more on RTMP and HTTP below).

- **Player.** Each current adaptive streaming system has a proprietary player, or similar component. Unless that player or component is installed on the target viewer's systems, he or she can't play the files.

With this as background, let's take a quick look at the technology alternatives.

Technology Alternatives

The adaptive streaming market is fast-moving with lots of players in and out. Here are the major players, roughly presented in the order of their entry into the market.

Move Networks

The first to popularize adaptive bit rate streaming in recent times was Move Networks, Inc., which had initial success with a range of high-profile sites that included ABC, FOX, ESPN, and Televisa. However, Move has seen its early technology advantage erode, lost most of its broadcast clients, and now has transitioned to selling its technology platform to Internet service providers that want to provide IPTV services.

What killed Move Networks? In "The Fall of Move Networks," *GigaOM* reporter Ryan Lawler found three reasons for the fall (bit.ly/fallofmove). First was the lack of installed base for the player, second the cost premium associated with the Move Networks technology, and third the inability of the technology to integrate with Flash-based advertisements that monetized the video.

Interestingly, in September 2010, Move Networks was awarded a "Fundamental Patent on Adaptive Video Streaming" (bit.ly/movepatent). Here's a blurb from the press release.

> The seminal patent covers the encoding and use of multiple bit rate "streamlets" with the same time index; such that the aggregate set of streamlets, stored on standard web servers, independently yield playback of the original full length video. This is accomplished through intelligent requests of successive unique optimal streamlets sent by a client device.

While you can't tell anything from the language in a press release, it seems like it would cover HTTP-based technologies like Apple's HTTP Live Streaming. Another blurb from the release states, "Following our market introduction in 2006, practically every major online video provider—including Akamai, Apple, Adobe, Limelight, Microsoft, Netflix, Widevine and several others—has deployed an adaptive bit rate architecture *inspired by* our invention" (emphasis supplied by author). The language inspired by is actually pretty curious, since "based-on" or "derived from" would have drawn a line in the sand, while "which infringes" would have had intellectual property attorneys salivating.

While it's unclear what Move intends to do with the patent, the Move Newsroom does point to a *NewTeeVee* article titled "Move Gets Streaming Patent; Are Adobe & Apple Hosed?" which you can find here: bit.ly/fFlkWd. Whatever its long-term intent, today Move Networks doesn't offer an adaptive streaming product that individual websites can implement.

Adobe: RTMP-Based Dynamic Streaming

Adobe has two adaptive streaming solutions, the first a Flash Media Server-based solution introduced in 2008. I'll get more into the religious battles surrounding HTTP vs. RTMP in the next section, but this solution requires the Flash Media Server (3.5 or higher) or equivalent (like the Wowza Media Server) and uses the RTMP protocol for distribution.

When choosing an adaptive streaming platform, player penetration is key. The ubiquitous, multi-platform Flash Player is obviously one of the strongest points of Adobe's product offering, since virtually all target viewers will have the player installed. RTMP-based Dynamic Streaming can also distribute video encoded with the VP6 and H.264 codecs.

Microsoft's Smooth Streaming for Silverlight

In late 2008, Microsoft announced Smooth Streaming for Silverlight, which was essentially the productization of the adaptive bit rate streaming technology originally deployed in the 2008 Summer Olympics. More recently, the 2010 Winter Olympics were streamed live via Smooth Streaming, along with NBC's Sunday Night Football offering.

As the name suggests, Smooth Streaming requires the Silverlight player, available on about 73% of all computers when I checked in April 2011. Smooth Streaming delivers via the HTTP protocol, and using that protocol has been the most widely touted competitive advantage over Adobe's RTMP-based Dynamic Streaming.

Apple's HTTP Live Streaming

In June 2009, Apple announced its own adaptive bit rate streaming technology, the afore-mentioned HTTP Live Streaming, which can be viewed on iDevices with cellular or Wi-Fi connections, or Mac computers running QuickTime X. This is the only technology that can be used to stream to iDevices and Android 3.0 devices, and it is primarily used in that role.

Akamai HD Network

Prior to the launch of Adobe's HTTP Dynamic Streaming offering, content delivery network Akamai launched Akamai HD Network, an HTTP-based service that can deliver adaptive streams to the Flash Player using HTTP on the Akamai network. The Akamai service uses "in the network" repackaging to input traditional RTMP-based streams from streaming produc-ers and convert them into chunks available for HTTP delivery to the Silverlight player or Apple iOS devices. Turner Broadcasting used this service to deliver its streaming presentation of the 2010 PGA Championship in August 2010.

Adobe HTTP Dynamic Streaming

In late 2010, Adobe launched HTTP-based Dynamic Streaming (originally code-named Zeri) to provide an Adobe-developed alternative for those seeking an HTTP-based technology. As the name suggests, the technology delivers to the Flash Player using the HTTP protocol, and doesn't require the Flash Media Server to operate.

At NAB 2011, Adobe announced support for HTTP Live Streaming via the Flash Media Server. Specifically, Adobe announced (adobe.ly/FlashdoesHLS):

> As we continue to evolve this technology we will be adding support for another format, HTTP Live Streaming (HLS). HLS is an MPEG2 transport stream (container) used by devices such as the Apple iPad 2. By adding support for HLS within the Flash Media Server, Adobe is reducing the publishing complexity for broadcasters who need to reach browsers supporting HLS through HTML5 (such as Safari) or devices

where Adobe Flash is not installed. Where Flash is installed, Flash Media Server packages the stream using MPEG4-fragments (F4F) to deliver video over HTTP to Flash.

There was no word on ship date in the blog post linked to above.

Scalable Video Coding (SVC)

Scalable Video Coding is an adaptive streaming extension to the H.264 standard. SVC's unique advantage is that it can serve multiple streams from a single encoded file, where technologies like Adaptive Streaming and Smooth Streaming need different streams for every supported configuration. For example, Major League Baseball streams 11 configurations within its baseball product offering, and has to create 11 unique streams for each game, plus backup streams for fail-over. This requires multiple racks of encoders, and obviously increases both storage and bandwidth cost.

With SVC, only one stream would be required (plus the fail-over), and it's only about 20% larger than the largest stream served. In addition, that single stream can dynamically adjust output resolution, data rate and frame rate to provide a much greater range of output streams. Where Adaptive Streaming requires 11 source streams to serve 11 different video configurations, a single SVC stream can serve as many as 48 different configurations, perhaps more. This makes SVC cheaper to encode, store and administrate, while providing a superior viewing experience to a broader range of viewers.

The problem with SVC is that it's currently unavailable, and unlike objects in your rear-view mirror, it isn't closer than it appears. The biggest stumbling block will be the player; to date, none of the major player vendors—Adobe, Apple and Microsoft—have announced support for SVC. Until this happens, SVC is a technology that looks great on paper, but hasn't yet become a real product. More on SVC at the end of this chapter.

Choosing a Technology

Now that you know the available technologies, how do you choose between them? Let's consider the two most relevant technology characteristics that most users are most likely to consider: supported platforms and protocol.

Supported Platforms

Probably the most important characteristic of any technology that you choose is whether it can reach your target viewers, either with or without a required player download (Table 7-1). Both flavors of Flash offer the broadest reach to desktops, and soon will extend to iDevices, with Smooth Streaming next at 73%. In contrast, Apple's HTTP Live Streaming cannot be played on Windows computers, and is only compatible with Macs running OS X version 10.6

and later (that percentage unknown), which means that it's not viable as a means for addressing desktop computers.

In terms of devices, Flash also has the lead in phone operating systems, with versions now available or soon to be available on Android, Palm, BlackBerry and others. Silverlight runs on the Windows Phone, which has only a small percentage of the market, while Apple's HTTP Live Streaming is compatible with all iDevices running iOS 3.0. While the percentage of iPhone and iPod touch devices running that OS is unknown, it's obviously a ton of devices, and beyond Flash, support for HTTP Live Streaming should be the most important priority for most streaming producers.

	Adobe RTMP-based Dynamic Streaming	Microsoft Smooth Streaming	Apple HTTP Live Streaming	Adobe HTTP-based Dynamic Streaming
Player	Flash	Silverlight	iOS 3.0 integrated Player/Safari/ QuickTime	Flash
Platforms	Win/Mac/Linux/ Solaris	Win/Mac/Linux	iOS 3.0/OS X 10.6 and later	Win/Mac/Linux/ Solaris
Player Penetration	96% +	73% +	unknown	96% +
Devices	Android, Palm, BlackBerry, others	Windows Phone	iDevices	Android, Palm, BlackBerry, others

Table 7-1. Supported platforms for adaptive streaming solutions.

Protocol (HTTP vs RTMP)

This argument gets to the heart of the differences between RTMP-based Flash and primarily Silverlight, and there's both a marketing and technology perspective. Let's start with the technology.

RTMP-based technologies require both a streaming server (like the Flash Media Server) and the player, and consistent communications between the player and server during the course of media playback. The connection is termed "stateful" because the player continually communicates its states via playback controls like play, pause or stop. In contrast, HTTP-based technologies don't require a streaming server—just the manifest file mentioned above and the player. The connection is called "stateless" because there is no persistent server/player communications.

These designations indicate a number of preliminary differences between the technologies. For example, regarding RTMP-based technologies, we know that:

- Since they require a steaming server, and servers cost money, they are likely more expensive.

- The stateful connection and server means that more playback information is captured, which likely means superior playback analytics.

There are several other key differences between the technologies, which include:

- HTTP is natively supported by caching devices on the web, which should make delivery more efficient than RTMP, which isn't natively supported by these devices.

- The availability of local caching should also improve quality of service for viewers served via the cache as opposed to those served by RTMP-based delivery, which is not likely to be cached.

- HTTP packets are not blocked by corporate firewalls, which should enable HTTP-based technologies to reach more viewers than RTMP-based technologies can.

To a degree—but only to a degree—most of these statements are true. We'll circle back in a moment, however, to determine whether the differences are relevant. Now let's discuss the marketing perspective.

I'm not sure what Microsoft's expectations were when it launched Silverlight, but it soon became clear that the platform was gaining little traction. Then Microsoft launched Smooth Streaming, which became the point of the spear for Silverlight marketing efforts. The primary difference between Smooth Streaming and Dynamic Streaming was the protocol; Smooth uses HTTP, while Dynamic uses RTMP, and Microsoft focused its marketing efforts on this distinction.

On its face, HTTP is "good" and RTMP is "bad." HTTP is the lingua franca of the web and an open standard, while RTMP is proprietary to Adobe. As mentioned, caching servers automatically cache HTTP data, but they have to be specially configured to cache RTMP data. HTTP data has no firewall issues, while RTMP packets may be blocked. If you listened to the Microsoft marketing speak, you would assume that RTMP-based Flash was a dead-as-Latin, legacy technology that cost much more than Silverlight and couldn't reach huge clumps of target viewers.

This sounded great in theory, but conflicted with the facts. For example, though RTMP-based Flash requires a server, single servers could handle thousands of simultaneous streams. In addition, because Flash is the most widely used technology, content delivery networks that deliver most high-volume events can amortize their server costs over multiple customers and multiple events. The bottom line is that few CDNs charge more for RTMP-based Flash delivery than they do for HTTP-based technologies.

The caching argument has also proven hard to quantify. To explain, video streamed via HTTP can be stored on cache servers located within the networks of ISPs, corporations and other organizations, while video streamed via RTMP cannot be cached without extra configuration that seldom occurs. These cache servers collect chunks of live or on-demand data that were requested multiple times and distribute them to multiple viewers, providing potentially higher-quality of service (QOS) to local viewers and reducing the overall bandwidth required to serve those viewers.

For example, imagine the Obama inauguration being watched by 20 viewers on a single network behind a cache server. Ten viewers are watching a stream from a news site distributing video via HTTP, and 10 are watching a stream from a site using RTMP. Since the HTTP data is cacheable, the website distributing via HTTP would send a single stream to the first requester. Once this stream was delivered, the cache server would cache and then serve the other nine viewers the cached stream. In contrast, the website distributing via RTMP would have to send 10 separate streams, increasing the overall transfer bandwidth and potentially delivering a lower QOS.

Again, these theoretical efficiencies have not translated into cost savings sufficient to convince large sites to convert from Flash to Silverlight. For example, MTV Networks, one of the largest distributors of video in the world, delivers its adaptive streams via RTMP-based Flash.

Regarding firewalls, Flash Media Servers can be configured to attempt several workarounds to deliver streams to viewers blocked by firewalls. In my conversations with streaming professionals, this doesn't seem to be a frequent occurrence. This impression is bolstered by the fact that the business-oriented *Wall Street Journal* delivers its Flash video via RTMP, which it obviously wouldn't do if firewalls prevented viewers in banks, investment firms and other corporations from watching the streams.

I'm not saying that Microsoft is lying, as HTTP does enjoy multiple theoretical benefits over RTMP—so much so that Adobe decided to deploy an HTTP-based streaming technology as well. What I am saying is that these theoretical benefits haven't translated into actual cost savings and that RTMP's theoretical disadvantages haven't dissuaded the vast majority of the streaming market to use RTMP-based Flash.

Now that HTTP-based Flash is available, it will be interesting to see how many current Flash users transition over to it. But for better or for worse, the availability of HTTP-based Flash also eliminates the clearest and most compelling technology advantage that Silverlight had over Flash—even if the advantages, in practice, never really panned out.

While there are multiple other factors to consider when choosing a technology, probably 99% of all technology decisions will be based on target platforms and protocols, so I'll move to the next major topic, which is how to support multiple adaptive streaming technologies.

Supporting Multiple Adaptive Streaming Technologies

Love it or hate it, HTTP Live Streaming is the only way to adaptively stream to iDevices, but it doesn't provide an adequate desktop solution unless post-2009 Mac computers are your only target. This means that most producers will have to support Flash OR Silverlight AND iDevices. In the past, this necessitated two essentially different encoding/delivery mechanisms to support streaming to the desktop and iDevices. Fortunately, that dynamic is changing, in large part because the Flash, Silverlight and iDevices can all play video encoded in the H.264 format.

One of the first solutions on the scene was from Wowza Media, which sells a streaming server that competes with Flash Media Server. As shown in Figure 7-3, the Wowza Media Server 2 can take one input stream and automatically convert it to all formats necessary to deliver to Flash (RTMP and HTTP), Silverlight, QuickTime and Apple iDevices. There's a very cool demo of that capability at www.wowzamedia.com/demos/demos.html.

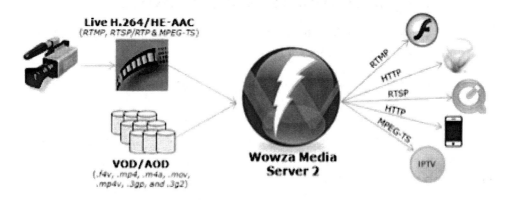

Figure 7-3. The Wowza Media Server can input one stream and deliver it to multiple devices using different technologies.

Essentially, Wowza is inputting an H.264 stream (or streams), re-wrapping it for the different container formats and protocols involved and creating any necessary manifest files. For example, while QuickTime, Flash and Silverlight most frequently work with files wrapped in the MPEG-4 container format, HTTP Live Streaming works with files encoded in the MPEG-2 container format. The Wowza server inputs the file in the MP4 container format, changes the header for the MPEG-2 container format, creates all the necessary file chunks and the manifest file, and delivers it to iDevices.

As mentioned, in September 2010, Akamai launched new features in its HD Network that performed a similar function. Here's a blurb from the press release.

> Our focus is to ensure each viewer gets the highest-quality experience possible leveraging the Akamai HD Network, while making it easier for customers to reach these growing platforms with more of their

content. We accomplish this through Akamai's "in the network" packaging. In a nutshell, companies provide Akamai with standard H.264 and MPEG4 video on-demand content, and we deliver it to iPhone, iPod touch and iPad. It's that simple.

The third alternative is from Microsoft, which provides the ability to re-wrap adaptive Silverlight streams for delivery to iDevices. Specifically, in Expression Encoder 4, you generate a Smooth Streaming MP4 bit stream and transmit that Microsoft's Internet Information Services 7 server with the IIS Live Smooth Streaming Feature. Like Wowza and Akamai, the server then re-wraps the transport stream from MP4 to MPEG-2, chunks the streams as necessary and creates the required .m3u8 manifest files that tell the iDevices playing the stream where to find and retrieve the video chunks.

Multiple format delivery is the future of adaptive streaming; while you can produce and deliver your desktop and iDevice streams separately, it makes little sense to do so.

Implementing Adaptive Streaming

With this as background, let's take a look at producing the files to distribute via the different adaptive streaming technologies, starting with Flash.

Adobe Flash Dynamic Streaming: RTMP

RTMP-based Dynamic Streaming works via communications between the Flash Player and Flash Media Server, with the Player monitoring factors like transfer data rate, buffer size and dropped frames to determine when to switch streams. It then requests data from an alternative stream, which the server delivers.

Separate files must be created for each stream, and the files are not chunked or otherwise specially prepared in any way. However, stream switches must occur on a key frame, which is why most authorities recommend using the same key frame interval for all streams.

We discussed encoding for H.264 in general in Chapter 4, so the terms in the following tables should be familiar. What I tried to do for this chapter was identify sources that detailed how to adjust standard encoding parameters for adaptive streaming. The problem is, there's no authoritative source that provides recommendations for all critical encoding parameters.

For this reason, Table 7-2 contains a collection of data points from these adaptive streaming-specific sources. I list them all, then present my recommendations on the extreme right, with discussions below each table. Note that not all sources have their own column in the table. Here are the sources I consulted for Dynamic Streaming with RTMP:

- Abhinav Kapoor, "Live dynamic streaming with Flash Media Server 3.5," (adobe.ly/kapoorlivefms) (Adobe1).

- Maxim Levkov, "Video encoding and transcoding recommendations for HTTP Dynamic Streaming on the Flash Platform," (adobe.ly/Levkovhttp) (Adobe2)

- Larry Bouthillier, "How to do Dynamic Streaming with Flash Media Server," *StreamingMedia* website (bit.ly/fmsdynamic)

- Case study data collected from MTV Networks and Harvard. Note that the author of the third article above, Larry Bouthillier, was the contact for the Harvard case study

- The adaptive streaming presets that Adobe released in late 2010 for Adobe Media Encoder, which you can read about at bit.ly/adaptivepresets.

- "Tutorial: On-demand HTTP Dynamic Streaming," Adobe website (bit.ly/ondemanddynamic)

- "Encoding Guidelines Dynamic Streaming for Flash over HTTP," Akamai website (bit.ly/akamaiwhitepaper). This paper is for Flash over HTTP, but all HTTP encoding recommendations should hold true for RTMP except for the key frame interval, which I discuss in more detail below.

	Adobe[1]	Akamai	Bouthillier	Harvard	MTV Networks	Recommended
How many streams	8	Not addressed	6	5	8	5-8
Resolution	All at least 4x4	All at least 4x4	All at least 4x4	All at least 4x4	All at least 4x4	At least 4x4
Data rate (kbps)	32, 80, 268, 468, 768, 1168, 1736, 2336	Formula based on motion in video	80, 192, 436, 672, 1072, 1672	64, 128, 256, 384, 768	400, 750, 1200, 1700, 2200, 3500	See Chapter 4
Frame rate	Not addressed	Not addressed	Full at high bit rates, 1/2 at low bit rates	Full unless under 150 kbps, then 1/2	Full	Full unless under 150 kbps, then 1/2
Profile/Level	Not addressed	Not addressed	Not addressed	Not addressed	Not addressed	Baseline to High

Table 7-2. Initial RTMP-based Flash Dynamic Streaming recommendations.

Here's some additional information on the key areas presented in Table 7-2.

How Many Streams?

In recommending eight streams, Adobe's Kapoor said the following:

> If your target viewer covers a broad spectrum of bandwidth capabilities, it is best to keep a wide range of stream bit rate encodings while keeping a large enough difference between successive bit rates. Too many bit rates too close to one another could result in too many stream switches, even with smaller bandwidth fluctuations. Besides the slight overhead in switching, the viewer's experience with too-frequent quality fluctuations may not be pleasant. Meanwhile, too few streams with a huge gap in bit rates would not provide the best quality or the most optimal stream for a particular bandwidth.

To summarize from this, you want:

- A wide range of streams to maximize the experience for all viewers
- That aren't too close together.

The only other factor that I'll add is the type of content being streamed and the target viewer, with entertainment-oriented sites tending to have more streams. You can see MTV in the Table 7-2 with eight streams, with Major League Baseball at 11 streams for its subscription-based service. In contrast, Harvard produces five streams, Indiana University (not shown in Table 7-2) produces three, and Deutsche Welle (a German news site, also not shown) streams four.

Overall, if you are producing advertising-supported entertainment, I would err on the side of more streams. If you're distributing information-based content and using adaptive streaming primarily to ensure a good experience for users at both ends of the connection/CPU spectrum, you can probably get away with fewer streams.

Video Resolution

The adaptive streaming sources included multiple recommendations regarding video resolution. These include:

- For the most part, none of our sources used exclusively 16x16 resolutions (called mod-16, explained in Chapter 5). Akamai included three mods—one for 16x16 resolutions, one for 8x8 resolutions and one for 4x4 resolutions—but preferred 16x16 over 8x8, 8x8 over 4x4, and 4x4 over any other resolution.

- Don't encode at resolutions larger than your source.

- When possible, try to match stream resolutions to the target playback window. MTV Networks conforms video resolutions to the most commonly used screen sizes on its web properties, and it never sends a video stream with a larger resolution than the playback window.

- If you must scale video to match a specific window size, scale up, not down. Scaling up uses less CPU than scaling down, which can slow playback by up to 40%.

In his white paper, Maxim Levkov points out that streams of identical resolution switch most smoothly. He recommends clustering streams at specific window sizes as shown in Figure 7-4. While there will inevitably be some switching between streams with different frame sizes, Levkov recommends using "ActionScript player logic to switch among targeted bit rates."

16x9				
Stream #	Picture Size	V	A	AV
1	256x144	150	64	214
2	256x144	250	64	314
3	512x288	450	64	514
4	512x288	600	64	664
5	512x288	800	64	864
6	512x288	1200	64	1264
7	768x432	1400	64	1464
8	1280x720	1700	64	1764
9	1280x720	2500	64	2564
10	1280x720	3500	64	3564

Figure 7-4. Adobe's Levkov recommends clustering your streams around certain frame sizes.

Clustering was the approach taken by Adobe with the streaming presets it released for the Adobe Media Encoder in late 2010, though I'd prefer a key frame interval that's a round number, like two or three seconds, rather than 72 frames (Table 7-3).

Preset	Resolution	Video bitrate	Profile	Key Frame	Audio bitrate	Channels
Phone and Table, 3G, 16x9	512x288	300 kbps	Baseline	72	48 kbps	Stereo
Phone and Table, Low, 16x9	512x288	450 kbps	Baseline	72	48 kbps	Stereo
Phone and Table, WiFi, 16x9	512x288	650 kbps	Baseline	72	48 kbps	Stereo
PC & TV, SD, Med, 16x9	768x432	1,140 kbps	Main	72	64 kbps	Stereo
PC & TV, SD, High, 16x9	768x432	1,140 kbps	Main	72	64 kbps	Stereo
PC & TV, HD, Low, 16x9	1280x720	2,440 kbps	High	72	64 kbps	Stereo
PC & TV, HD, High, 16x9	1280x720	3,440 kbps	High	72	64 kbps	Stereo

Table 7-3. The configurations of adaptive streaming presets released by Adobe in late 2010.

MTV Networks was kind enough to share its stream configurations, which I show in Figure 7-5. As you can see, other than the two lowest-quality streams, which are targeted at mobile and other constrained devices, MTV configured its videos to fit specific screen sizes, and all streams have different resolutions.

As mentioned above, MTV configured its player logic so that the stream sent to the player will never have a larger resolution than the current playback window. For example, if the viewer were watching within a medium in-page window (640x360), MTV would never send a higher-quality stream to that viewer unless her or she increased the size of the playback window. That makes wonderful sense, because scaling downward can affect player performance, and since the quality of the streams are close to identical (as measured by the bits-per-pixel values), the viewer wouldn't notice the difference anyway.

Scenario	Format	Frame Size	Total Bitrate	Audio Bitrate	bits/pixel *frame @ 30 fps	bits/pixel *frame @ 24 fps
Mobile & constrained (low)	baseline, mono, 10 fps	448x252	150	48	0.09	0.09
Mobile & constrained (high)	baseline, mono	448x252	450	48	0.12	0.15
Sidebar placements	main profile, stereo	384x216	400	96	0.12	0.15
Small in-page	main profile, stereo	512x288	750	96	0.15	0.18
Medium in-page	main profile, stereo	640x360	1200	96	0.16	0.20
Large in-page	main profile, stereo	768x432	1700	96	0.16	0.20
Full size in-page	main profile, stereo	960x540	2200	96	0.14	0.17
HD 720p (full screen)	high profile, stereo	1280x720	3500	96	0.12	0.15

Figure 7-5. MTV's encoding schema targets video resolutions to specific window sizes. Image courtesy MTV.

Note that upon the initial connection, MTV directs the viewer into one of two buckets: mobile and constrained, which are the top two streams in Figure 7-5; and broadband, which are all others. While viewers can adaptively switch streams within each bucket, the current MTV schema doesn't let them switch buckets. Finally, as the name suggests, the mobile and constrained feeds are the ones pushed to Akamai for "in the network" remuxing for delivery to iDevices.

To summarize, if you're working with three or four different window sizes, consider creating several adaptive streams for each window size. If your technology enables this, you should make sure that the viewer never gets to switch to a stream that's larger in resolution than the current viewing window. Obviously, if you have to support seven or eight display sizes, like MTV, you should make sure that you have at least one stream for each window size.

Video Data Rates

Here's what the different sources had to say about video data rates:

Adobe's Kapoor: Keeping too many bit rates close to one another could result in too many stream switches, even with smaller bandwidth fluctuations. Besides your (slight) overhead in switching, the viewer's experience with frequent quality fluctuations may not be pleasant. On the other hand, encoding too few streams with a huge gap in bit rates would not provide the best quality or the most optimal stream for a particular bandwidth environment.

Bouthillier: Selection of available bit rates: The most successful tests I conducted include a number of streams with fairly close-together bit rates. When network conditions fluctuated, a stream switch that involved a large jump in bit rate or screen size was distracting. Stream switches between more similar bit rates were often barely noticeable and provided a more pleasant user experience.

One other factor in choosing video data rates is maintaining a consistent quality level over all streams. In this regard, Chapter 4 details how to use the bits-per-pixel value to maintain the quality of the individual streams.

Another approach is taken by Microsoft's Alex Zambelli, who created a Smooth Streaming Multi-Bit-Rate Calculator (alexzambelli.com/WMV/MBRCalc.html) which is shown in Figure 7-6. Operation is simple: You fill in the information on the top of the screen—Max Width, Aspect Ratio and the like—and then press Go. The calculator then suggests the streams shown in the display box, complete with recommended bit rate and resolution, and aspect ratio data.

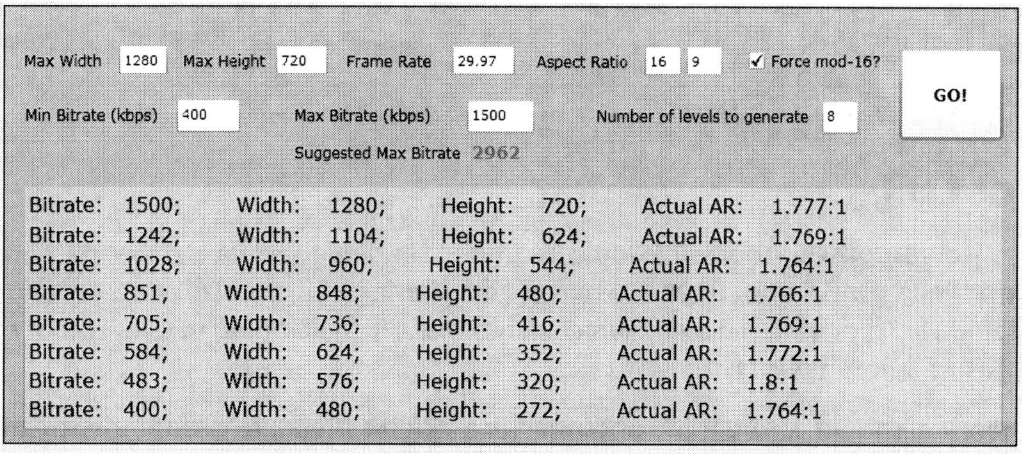

Max Width	1280	Max Height	720	Frame Rate	29.97	Aspect Ratio	16	9	✓ Force mod-16?		GO!
Min Bitrate (kbps)	400			Max Bitrate (kbps)	1500		Number of levels to generate	8			

Suggested Max Bitrate **2962**

```
Bitrate:  1500;   Width:  1280;   Height:   720;   Actual AR:  1.777:1
Bitrate:  1242;   Width:  1104;   Height:   624;   Actual AR:  1.769:1
Bitrate:  1028;   Width:   960;   Height:   544;   Actual AR:  1.764:1
Bitrate:   851;   Width:   848;   Height:   480;   Actual AR:  1.766:1
Bitrate:   705;   Width:   736;   Height:   416;   Actual AR:  1.769:1
Bitrate:   584;   Width:   624;   Height:   352;   Actual AR:  1.772:1
Bitrate:   483;   Width:   576;   Height:   320;   Actual AR:  1.8:1
Bitrate:   400;   Width:   480;   Height:   272;   Actual AR:  1.764:1
```

Figure 7-6. Alex Zambelli's Smooth Streaming calculator.

According to Zambelli, to compute the values, "the resolutions and bit rates are plotted against a power curve that approximates the relationship between bit rate, resolution and quantization. In other words, we try to keep the quantizer parameter roughly consistent for all bit rates in order to ensure consistent compression quality." Though it's called the "Smooth Streaming" calculator, there's no reason that you shouldn't at least consult it for other adaptive bit rate technologies—of course, adjusting the results for other inputs as discussed above.

Profile and Level

If you're producing solely for computer playback, use the High profile for all encodes. If you're producing a stream that will be transmuxed and delivered to iDevices, you must produce streams compatible with your targets, which means Baseline/Level 3.0 for pre-4G/iPod touch devices, and Main/Level 3.1 for the iPad and 4G iPhone/iPod touch devices. Check the Apple/Akamai recommendations below for more details.

Though it's not listed in any table, I would enable CABAC for all encodes where it was available. If you're producing for computers using the High profile, enable it; If you're producing for the iDevices with the Baseline profile, it's unavailable.

	Adobe[1]	Akamai	Bouthillier	Harvard	MTV Networks	Recommended
VBR/CBR	Not addressed	Not addressed	CBR	CBR except for highest	VBR, 2x constrained	CBR
Key frame interval	5 seconds, consistent	2-4 seconds, consistent	2 seconds, consistent	2 seconds, consistent	2 seconds, consistent	2-5 seconds, consistent
Client side buffer	6-10 seconds (larger than key frame interval)	Not addressed	2s key frame interval	2x key frame interval	Not addressed	2x key frame interval
Audio parameters	Same in all files	Not addressed	varied 16-128 kbps	varied, 32-128 kbps	96 kbps, consistent	vary with stream, but test

Table 7-4. Final RTMP-based Flash Dynamic Streaming recommendations.

VBR or CBR

The knee-jerk response to this issue would be CBR to simplify the data transfer, particularly over constrained connections. That is, when encoded using VBR, the size of a 5-second file chunk could vary pretty dramatically, which could make it challenging to deliver that chunk in time to ensure smooth playback.

In addition, you should consider the potential for a VBR bit stream to disrupt the stream-switching algorithms implemented by the adaptive streaming schema. Specifically, buffer status is one of the factors monitored to determine if a stream switch is necessary. If a high-motion scene within a video was encoded at 2x the normal data rate, during transmission to the viewer, which would take longer than normal, the buffer could drop below a threshold level, triggering a stream switch that would not have been necessary with a CBR stream.

On the other hand, if MTV can use 2x-constrained VBR, perhaps you can too. Harvard presents a nice compromise, which is using CBR for low-bit-rate streams, and VBR for higher-bit-rate streams. Certainly the most conservative approach would be to use CBR for all streams, which was the approach used by Adobe with their Adobe Media Encoder presets and what I recommending. If you do try VBR to boost the quality of your streams, make sure it's constrained to no higher than 2x, and consider using VBR only on the highest-bit-rate streams in your adaptive streaming package.

Key Frame Interval

As discussed above, stream switching must occur on a key frame, so use the same key frame interval in all encoded streams. All resources and users reported that the typical 10-second interval was just too long, with Adobe recommending between 2 and 5 seconds and most users in the 2-second range, hence the 2-5 second recommendation.

One big question is whether to enable scene change detection, which obviously inserts key frames at scene changes, then restarts the regular key frame insertion interval. So long as you're encoding all streams with the same encoder, the logic shouldn't change, and the program should insert a key frame at the same frame in each stream, which enables seamless switching between streams.

On the other hand, the most conservative view would be to disable scene change detection, absolutely ensuring that key frames are inserted at the same location in each stream. So by "consistent" in the table, I mean disable scene change detection.

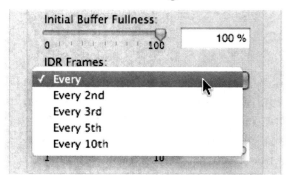

Figure 7-7. Make all I-frames IDR frames when encoding for adaptive streaming.

One key frame-related option available in Telestream Episode, and most enterprise encoding tools like Rhozet Carbon Coder and Inlet Fathom, involves the concept of the IDR frame, as shown in Figure 7-7 (You can view the complete discussion of <u>IDR Frames</u> in Chapter 4). Here's how the Episode help file describes IDR frames:

> IDR Frames-An IDR frame is an I-frame whose preceding frames cannot be used by predictive frames. More distant IDR frames may allow more efficient compression but limits the ability of a player to move to arbitrary points in the video. In particular, QuickTime Player may show image artefacts when you scrub the timeline unless every I-frame is an IDR frame.

According to Adobe's Levkov, "Flash can only change bit rates at IDR frame intervals." He therefore recommends (I think; it's a pretty dense discussion) that all key frames be IDR frames. I recommend the same (I'm sure); so if the option shown in Figure 7-7 is available, make every key frame an IDR frame. Note that if your encoding tool doesn't specify IDR frames, they almost certainly are (so don't worry about it).

Note that key frame intervals tend to be shorter with HTTP streaming than with RTMP. This is the RTMP section, with HTTP for Flash discussed below, so I'm comfortable recommending an interval of up to 5 seconds. For HTTP streaming, I'd be more inclined to use the low number in Akamai's recommendation, which is 2 seconds. More on that below.

Audio Parameters

The first Adobe white paper recommends keeping "the audio bit rates and sampling rates the same to provide a seamless switch between them" and warns that switching between incompatible bit rates or sampling rates might result in an audible "pop" at the transition. However, *in that same white paper*, Adobe presents a range of audio parameters, from 16 kbps mono to 128 kbps stereo. The paper also recommends that if you do offer different streams, switch from stereo to mono using the same per channel sampling rate and bit rate. For example, if your stereo stream is 128 kbps (44,000 sample rate, 16-bit sample size), use a mono stream that's 64 kbps with the same sample rate and sample size.

In the presets that Adobe released for the Adobe Media Encoder, the lowest bandwidth streams used 48 kbps stereo, while the higher bit rate streams used 64 kbps stereo. I guess the takeaway from Adobe would be use different audio configurations, but also to test to ensure that there is no popping or other issues when switching streams.

There was diversity in actual user practices, with MTV consistent with a single 96 kbps audio stream, but Harvard matching adjusting audio bit rate with the video bit rate. Of course, the bit rate of Harvard's lowest-quality video streams (64, 128, 256, 384 kbps) were substantially lower than MTV's (400 kbps), which was by design, because many users watch Harvard's videos at these lower data rates. Matching a 96 kbps stereo audio stream with a 64 kbps video stream makes little sense, which is why Harvard dropped to 32 kbps for these streams.

Overall, though you'd have to test to be sure that you could switch streams without audible popping, I recommend a hybrid approach. For higher-quality video streams, use a single high-quality audio stream (say, 96 kbps stereo). For lower-quality streams sent to viewers on low-power devices or constrained connections, use a lower-quality stream (say 32 kbps).

Adobe Flash Dynamic Streaming: HTTP

Let's take a quick look at Adobe's HTTP-based Dynamic Streaming here since most of the encoding recommendations for RTMP-based streaming will stay the same. Figure 7-8 presents an overview of both the on-demand (VOD) and live workflows.

Let's start by looking at the File Packager phase in the Preparation step. Like most HTTP-based technologies, Adobe divides the encoded files into fragments (files with the .f4f extension) and creates a manifest file (.f4m extension) that the player will use to locate the alternate bit rate streams and request the relevant fragment.

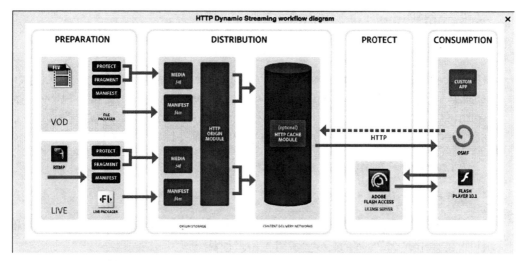

Figure 7-8. Adobe's HTTP Dynamic Streaming workflow diagram. Image courtesy Adobe.

The files then get transferred to the HTTP Cache Module, with the option to protect the streams via the Adobe Flash Access License Server. Then the streams are consumed via either a custom app, an Open Source Media Framework (OSMF) player or the Flash Player itself.

I would use the guidelines discussed in Tables 7-2 and 7-4 as starting points, as modified by these key differences between encoding for RTMP-based Dynamic Streaming and for HTTP-based Dynamic Streaming.

- **Regarding I-frames.** Adobe's Levkov recommends using an I-frame interval of 3-8 seconds for long form content (1 minute or longer), with all I-frames designated as IDR frames, which I discuss above. I haven't looked at every tool used to convert encoded files into fragments, but the Adobe command line File Packager uses the key frame interval of the encoded file (among other characteristics) to determine fragment length, ensuring that all fragments start on a key frame. If your encoding tool doesn't let you specify which I-frames are IDR frames, all I-frames are likely IDR frames (so don't worry about it), and if it does, just make sure that all key frames are IDR frames.

- **Also regarding key frames.** Akamai recommends an interval of 2-4 seconds. I would start with 4 seconds and shorten the interval to no less than 2 seconds if 4 seconds proves unresponsive.

- **Also regarding key frames.** Levkov recommends disabling natural or key frames at scene changes. This ensures that all key frames are at a regular interval. Levkov also made the following recommendations, though the white paper is a little light on supporting detail for my taste:

- B-frame interval of 2 maximum.

- 2-4 reference frames, "closer to 2."

- Two-pass CBR recommended when available (i.e. when not live). Levkov does state, "VBR bit rate maybe used as well, however, exceeding 10% above the minimum set bit rate is not recommended." I read this to say that you can use constrained VBR with the maximum set to 10% over the target.

- A consistent frame rate across all files.

- Buffer length of 1-2 seconds.

- For encoding applications that give you access to the Video Buffer Verifier (VBV), buffer of 70% initially, and 100% final.

- In very detailed tables at the end of the white paper (adobe.ly/Levkovhttp), Levkov recommends using 16-bit, 44 kHz stereo audio encoded at 64 kbps for all encoded streams. These tables also recommend using the Main and High profile, though exclusively CAVLC rather than CABAC.

Levkov's work is the most detailed that I've seen, and it certainly should be reviewed by anyone implementing HTTP Dynamic Streaming. In my view, the CAVLC decision may be unduly conservative and could unnecessarily subdue quality, but conservative is probably a good approach for all producers implementing HTTP-based adaptive streaming for the first time.

Apple's HTTP Live Streaming

Let's transition over to Apple's HTTP Live Streaming—which can be used to adaptively stream to both iDevices and Android 3.0 devices—first getting an overview and then digging into the details. For the overview, Figure 7-9 is from an Akamai White Paper on HTTP Live Streaming. On the bottom left, you see the video input into the system where it's encoded and divided into chunks in the segmenter. Not shown is the creation of the M3U8 index files, which identify the alternate streams and the location of all data chunks.

The index and video files can be hosted on the customer's own web server or in Akamai Net Storage, and then they are distributed through the Akamai HTTP Network (if you use Akamai as your CDN). iDevices connecting via Wi-Fi get delivery direct from the Net, while those connecting via wireless get the data through their wireless carrier.

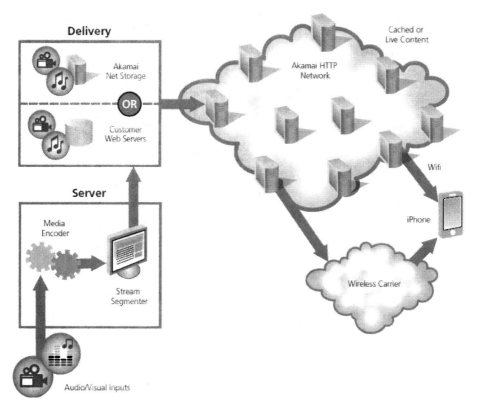

Figure 7-9. Overview of Apple's HTTP Live Streaming with Akamai as the CDN. Image courtesy Akamai.

With this as background, let's focus on encoding parameters. To create Table 7-5, I reviewed the following resources.

• Apple Tech Note: "Best Practices for Creating and Deploying HTTP Live Streaming Media for the iPhone and iPad," (**bit.ly/bestpracticehttplive**)

• Apple Tech Note: "HTTP Live Streaming Overview," (**bit.ly/httpliveoverview**)

• Apple Tech Note: "Using HTTP Live Streaming," (**bit.ly/usinghttplive**)

• Akamai White Paper: "Akamai HTTP Streaming for iPhone Best Practices," (**www.akamai.com/html/perspectives**)

• Case study data collected from Turner Broadcasting, MTV Networks, Harvard and Deutsche Welle.

To be clear, the starting point for anyone encoding for HTTP Live Streaming should be the Best Practices Apple Tech Note, which supplies most of the relevant data. I tried to fill in the small remaining major gaps in Table 7-5.

	Apple Best Practices Tech Note	Harvard	Turner	Deutsche Welle	MTV Networks	Recommend
Resolution	400x224, 640x360	320x240, 400x300	400x224	640x360, 400x224	448x252	Apple for dedicated iDevice
Data rate (kbps)	100-1.2, 1.6 max	132, 564	126, 450, 800	100, 200, 400, 800	150, 450	Apple
Frame rate	10, 15, 29.97	15, 29.97	10, 29.97	8, 12.5, 25	native	Apple
Profile/Level	Baseline/3.0 to Main/3.1	Not addressed	Not addressed	Not addressed	Baseline	conform to target
VBR/CBR	Not addressed	CBR	Not addressed	CBR	VBR 2x constrained	CBR
Key frame interval	3 seconds, consistent	2 seconds, consistent	2 seconds/ 3 seconds	50 frames, consistent	1 second/ 2 seconds	2 seconds, consistent
Segment duration	10 seconds	Not addressed	Not addressed	Not addressed	Not addressed	10 seconds
Audio parameters	40 kbps, 22.05 sample rate, consistent	32 kbps/ 64 kbps	16 kbps/ 40 kbps	64 kbps, consistent	48 kbps/22 kHz mono	48 kbps/22 kHz/mono, consistent

Table 7-5. Summary of HTTP Live recommendations.

Let's discuss some of the major items.

Resolution

Regarding resolution, Turner and Deutsche Welle, which were both producing streams solely for iDevices, followed Apple's recommendations, while Harvard and MTV, which were repurposing streams created for Flash, did not. For the record, Harvard was using the Wowza server to transmux the streams, while MTV Networks was using Akamai's "in the network" repackaging. If you are producing streams specifically for iDevices, I would follow Apple's recommendations. If you're repurposing other content, you'll have to find the optimum configurations for both roles, perhaps using Harvard and MTV as starting points.

Profile/Level

As mentioned, if you're directing the same stream to computers and iDevices, the profile/level that you choose must be playable on the lowest targeted iDevice platform, which means Baseline. Interestingly, you can mix streams encoded in the Main and Baseline profiles in the same HTTP Live presentation. As mentioned, the index file contains a list of alternate streams and the codecs required to play the stream. If you're watching from a pre-4G iPhone or iPod touch, the device will check the index file for alternative streams, automatically ignoring any that require a Main-profile-compatible player to play back.

VBR/CBR

CBR is the most conservative approach, particularly since you're distributing either over cellular or Wi-Fi, never a cabled connection.

Key Frame Interval/Segment Duration

Though there is some static in other directions, using a key frame interval of 2-3 seconds consistently in all streams is again the most conservative approach. Remember, however, to consider the duration of the media segment when choosing a key frame. Here's a blurb from the Wowza Media Segmenter documentation that provides specific direction:

> Chunks must start on a key frame. So it is best to use a key frame interval that is factor of the chunkDurationTarget setting. For instance if chunkDurationTarget is set to 10 seconds then use a key frame interval of either 2, 2.5, 5, or 10 seconds.

In non-Wowza-speak, make sure that whatever key frame interval you choose, it divides evenly into the selected segment size. Here's a blurb from the above referenced Akamai white paper: "Key frames are suggested every 5 seconds, ideally an even divisor of the chosen segment length." The white paper goes on to say, "Apple and Akamai both recommend 10-second segments as the best length for balancing performance and user experience. Shorter chunk sizes will incur additional transfer overhead and result in more file entries."

The reason I'm belaboring this seemingly obvious point is that Apple's HTTP Live Streaming Overview recommends a segment interval of 10 in FAQ 3, page 21, and a key frame interval of 3 in FAQ 10, page 23. Obviously, 3 doesn't divide evenly into 10. There's probably a good explanation for this, but it's not immediately apparent to me.

To be on the safe side, make sure your key frame interval divides evenly into your segment length. If you're looking for an absolute recommendation, I'd recommend a key frame interval of 2 seconds, and a segment length of 10 seconds.

Now onto Silverlight.

Smooth Streaming with Silverlight

Here's a list of sources that I consulted regarding my recommendations:

- "Akamai HD for Microsoft Silverlight On-Demand Encoding Recommendations," (bit.ly/akamaiwhitepapers)

- Alex Zambelli, "IIS Smooth Streaming Technical Overview," (bit.ly/smoothstreamzambelli

- Case study data collected from Microsoft regarding the Olympics Broadcast and NBC's Sunday Night Football broadcast.

The two case studies involved VC-1 video, not H.264, and the other white papers are either very basic or don't offer specific advice regarding producing H.264 video for Smooth Streaming. Fortunately, Microsoft's Expression Encoder 4 is very specific regarding H.264 parameters for Smooth Streaming, and that's the bulk of the data contained in Table 7-6.

	Olympics	Sunday Night Football	Expression Encoder	Recommend
How many streams	6 per input, 8 total configurations	7 for HD input, 6 for SD input	10 for 1080p, 8 for all others	6-8
Resolution	Most are mod-16, all at least mod-8	2 are mod-4, all others mod-16	Most mod-16, all at least mod-4	At least mod-4, try for mod-16 in smaller sizes
Data rate (kbps)	400, 600, 900, 950, 1250, 1600, 1950, 3450	50, 350, 600, 950, 1500, 2250, 3450	230, 331, 477, 688, 991, 1400, 2100, 3000, 5000, 6000	Smooth Streaming Calculator
Frame rate	29.97 for all streams	29.97 except lowest-quality stream (15 fps)	All match source	Source
Profile/Level	NA	NA	Main	Main
Entropy encoding	NA	NA	CABAC	CABAC
B-frame interval	1	1	1	1
Reference frames	Unknown	Unknown	4	4
VBR/CBR	CBR for live, 110% constrained VBR, on demand	All live, all CBR	CBR-1 pass	CBR
Key frame interval/chunk size	2 seconds	2 seconds	2 seconds	2 seconds
Client side buffer	5 seconds	5 seconds	4 seconds	4 seconds
Audio parameters	48 kbps for all streams	48 kbps for all streams	160kbps stereo for all streams	48 kbps mono for all streams

Table 7-6. Summary of Smooth Streaming for Silverlight recommendations.

Most of the recommendations in the table are self-explanatory, but I wanted to share some points made by Microsoft's Alex Zambelli in a *StreamingMedia* webinar on March 24, 2011. You may have to register to get them, but his slides (and those used by other speakers) should be available online at bit.ly/smtranscoding. Here are the major points.

- **Use only a single audio bit rate.** I didn't know this, but Zambelli stated that the Silverlight client only supports a single audio bit rate with Smooth Streaming, so don't change audio parameters when you're producing your streams.

- **Use a single resolution if streaming to Window Phone 7.** Didn't know this either, but Windows Phone 7 cannot switch streams with different resolutions, so if you're streaming to a Windows Phone, use a single resolution for all streams.

- **Use a 2-second key frame interval.** Confirming my recommendation.

- **Disable scene-change detection.** No surprise here, but Zambelli noted that all key frames must be aligned across the files, so you should disable scene-change detection, or the equivalent feature, in your encoding program.

- **Set client side buffer at 2-3x key frame interval.** Table 7-6 recommends 4 seconds, which is 2x the recommended 2-second key frame interval.

- **Drop to 15 fps at lower than 300 kbps rather than reducing the resolution.** Zambelli said that this delivered superior quality over streaming at a lower resolution.

Zambelli also confirmed that Microsoft's IIS Media Services 4.0 can dynamically transmux live Smooth Streaming files supplied by Expression Encoder 4 into Apple's HTTP Live Streaming format for delivery to iOS and Android devices. The obvious common codecs are H.264 and AAC-LC. Note that if you're also delivering to Windows Phone 7 devices, you cannot change the stream resolution, nor can you change the audio configuration if you're delivering to Silverlight.

Otherwise, the remainder of the recommendations in Table 7-6 speak for themselves.

Multiple Format Distribution

Of course, if you're originally creating Flash or Silverlight H.264-based streams and are using a transmuxing technology to re-wrap the files for delivery to iDevices, you need to meld the recommendations to fit all target platforms. In particular, mind Apple's recommended target resolution for older iDevices (400x224) and that all pre-4G iPhones and iPods can only play video encoded using the Baseline profile.

Scalable Video Coding

Before you select an adaptive streaming technology, let's take a deeper dive into Scalable Video Coding (SVC), which is an H.264 extension for adaptive streaming. Rather than producing multiple files, the SVC encoder converts the source video into layers, starting with the Base layer, which contains the lowest level of quality and is backward-compatible with existing H.264 players. In addition to the Base layer, the encoder creates Clarity layers, which contain additional frame-based detail, and Size layers which contain additional resolution.

For example, the Base layer of an encoded stream might be encoded at 320x240 resolution with a data rate of 150 kbps for delivery to iPhones. Additional layers could expand that stream to 720p video at 3 Mbps to produce a stream suitable for delivery to a set-top box.

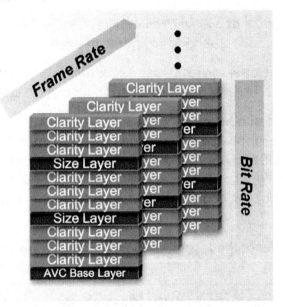

Figure 7-10. The SVC stream is composed of Base, Clarity and Size layers (image courtesy Seawell Networks).

Unlike existing technologies that produce each stream separately, when it's creating the SVC stream, the encoder incorporates all layers into a single file, roughly 20% larger than the size of the largest stream. In our example, the data rate of that file would be 3.6 Mbps, much smaller than the combined size of the multiple separate files created by alternative technologies. This reduces transfer and storage costs over other adaptive streaming technologies, and because a single SVC encoder can create a stream that serves multiple targets, it eliminates the need to purchase multiple encoders.

The other special sauce of the SVC file is the ability to juggle frame rate, size and clarity to produce streams in a wide variety of configurations (Figure 7-11). That is, when the viewer clicks the link for the SVC-encoded video, the player retrieves the Base layer plus additional layers necessary to create the optimal playback experience. An iPhone might retrieve only the Base layer, a Wi-Fi connected netbook might retrieve the Base layer plus several Clarity and Size layers, and the set-top box would retrieve the entire file. As we'll see in a moment, the first commercial SVC hardware encoder can reportedly produce a single file that delivers 27 or more playback configurations. This flexibility enables much greater playback optimization than alternative technologies working with discrete encoded files.

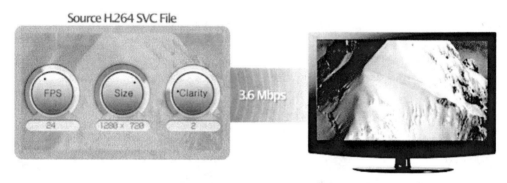

Figure 7-11. SVC streams can juggle frame rate, size and clarity for 48 playback configurations (image courtesy Seawell Networks).

To summarize, SVC cuts initial capital investment over existing technologies because a single encoder can produce a steam that serves multiple configurations. Because the SVC stream is smaller than the multiple separate streams created by existing technologies, SVC-based video streams are less costly to transmit, store and administer. Because of the layer-based architecture, a single SVC stream can create a broad range of playback configurations, optimizing the experience for a wider variety of platforms and connection speeds. Finally, because it's an open standard, SVC should be more palatable to all stakeholders, improving its chance of broad-based adoption.

Available When?

When will H.264 SVC become generally available? Within the context of widespread adoption, it's going to be a while, because all three elements—encoder, server and player—must be SVC-aware to fully leverage the technology. CDN support will also be necessary for larger outlets.

The wheels are in motion, however. For example, H.264 encoding vendor MainConcept showed a technology preview at IBC 2008 in Amsterdam, and it is currently looking for technology partners for the end-to-end technology components—encoder, decoder and network components. However, the technology will likely find its initial implementations in closed systems, like the Google's Gmail Chat, which is based upon H.264 SVC technology licensed from Vidyo (blogs.zdnet.com/Google/?p=1176), and security applications like those offered by GE Security (bit.ly/h264svcge).

In May 2010, Toronto-based Seawell Networks, which supplied the figures shown in this section, announced the Lumen 1000, the first Scalable Video Coding encoding appliance. Here's a blurb and picture from the press release:

> The Lumen 1000 creates H.264 SVC files that contain multiple resolutions and deliver only the bits required for playback on a given device, at a given network speed. The Lumen 1000 can generate SVC files that contain 27+ resolution layers, giving content owners more control over their media than ever before.

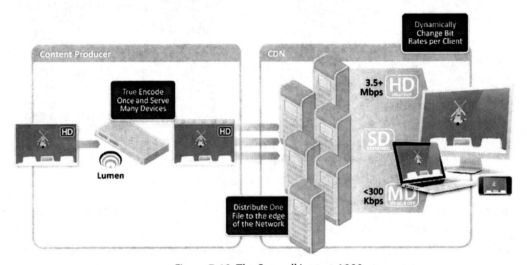

Figure 7-12. The Seawell Lumen 1000.

On the other hand, at NAB 2011, Seawell introduced Spectrum, with this description on their website:

> Spectrum is a high performance application that can instantly repackage live or on-demand video to support all adaptive streaming protocols (such as Microsoft Smooth Streaming, Apple HLS, and Adobe HTTP Dynamic Streaming) and ease the introduction of new protocols in the future. By repackaging the content "on-the-fly", video content is only repurposed as needed for distribution to clients, which drastically decreases storage and distribution costs.

In other words, it's a transmux-type technology, with no mention of SVC in sight. I'm a believer in SVC technology, but Seawell hasn't announced that anyone's actually distributing SVC streams yet, and no major player vendor—like Adobe, Apple or Microsoft—has announced support for SVC playback, which I think is essential for the technology to become pervasive. That said, standards-based technologies can gain tremendous momentum very quickly, so no one should make an adaptive streaming technology decision without at least checking the status of SVC.

That's it for adaptive streaming; in the next chapter, we'll look at choosing an H.264 encoder.

Chapter 8: Choosing an H.264 Encoding Tool

By now, you know pretty much everything you need to know about producing H.264 files, at least in the abstract. Where the rubber meets the road, however, is when you convert theory to practice and actually produce an H.264-encoded file. This obviously takes an H.264 encoding tool.

You should consider at least four elements when choosing an encoder to produce H.264 video files. Most important is video quality, and though the playing field has largely leveled out here, there still are some discrepancies—most notably with Apple Compressor using the Apple H.264 codec. Simply stated, it's subpar.

Assuming quality is similar, your next focus should be encoding speed, particularly in high-volume operations. Next up are production-oriented features that expand the basic utility of the encoding tool. For example, the ability to retrieve and encode files from remote FTP sites can be very useful, as is watch folder functionality, which allows any network user with access to a shared folder to encode a file to a predefined preset. Some tools can also accelerate encoding via cluster operation, allowing one user to create jobs that are encoded on multiple computers.

Finally, some encoding tools offer features like the ability to add intro and outro videos to the start and end of the video, which is useful for marketing and branding, or add a watermark or logo to the video. I'll examine all these features for a range of commercial encoding tools for both Mac and Windows platforms.

Mac Encoders

Let's start with the Mac. Table 8-1 shows the basics of Mac (and cross-platform) standalone encoding programs. Note that all encoders other than Compressor use the MainConcept H.264 codec, which enjoys near universal use in non-Apple commercial encoding tools. In addition, you'll see that Compressor doesn't natively produce VP6 or WMV, so if you need to produce those formats and are seeking a complete solution, Compressor isn't it. Adobe Media Encoder does support VP6, but produces WMV only in the Windows version, not on the Mac.

	Adobe Media Encoder	Apple Compressor	Apple Compressor with x264Encoder	Sorenson Squeeze	Telestream Episode Pro 6
Platforms	Mac/Windows	Mac	Mac	Mac/Windows	Mac/Windows
Price as tested	Bundle-only	Bundle-only	Free	$799	$995
H.264 codec	MainConcept	Apple	x264	MainConcept	MainConcept
Other streaming codecs supported	VP6	None	None	VP6, WMV	VP6, WMV

Table 8-1. Mac encoding tools.

Now let's spend some time getting to know each encoder.

Apple Compressor

I encoded twice with Compressor, once using Apple's own H.264 codec, and once using the x264Encoder plug-in, which you can download for free at bit.ly/X264encoder (I used x264Encoder 1.2.19, dated 12/8/2010). Basically, Apple's codec works well at higher-than-average data rates (Apple encoded its iPhone 4 video marketing video at a very high .26 bits per pixel, presumably with Compressor) but it is very uncompetitive for those seeking the highest possible quality at the lowest possible data rate. You can read more about that in "x264Encoder vs the Apple Codec" at bit.ly/x264vApple. With both codecs, I encoded with Qmaster in all trials (you will learn how to set up Qmaster in Chapter 12).

I used the settings shown in Figure 8-1 when encoding with the Apple H.264 codec. Note that all files encoded via this configuration use the Main profile—to produce files with the Baseline profile, you have to use an Apple iDevice preset. This configuration also produced files using CAVLC entropy encoding, rather than the higher-quality CABAC, and all files had a B-frame interval of one (IBPBPBPBPB ... and so on). To produce files without B-frames, uncheck the Frame Reordering checkbox.

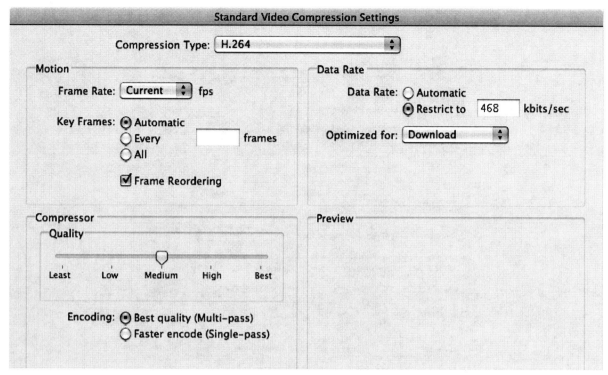

Figure 8-1. Settings used in Compressor with the Apple codec.

I used the Automatic Key Frames option because *Apple Technical Note TN228* states, "For H.264, we recommend leaving the key frame interval up to the compressor; so you should choose 'Automatic' for the best quality result." With Automatic selected, the encoder inserted key frames every 90 frames, and it did insert key frames at scene changes.

When producing the comparison test file with x264Encoder, I used the settings shown in Figure 8-2. To open this window, click the Options button that you'll see in the Standard Video Compression Settings screen after you download and install the codec. Then, click the Load preset option on the lower left of the libavcodec settings dialog. Click the Use Library native preset/tune radio button, and choose a preset. The notes to the encoder tell you to use Slow, Medium or Fast, and I used Slow without any customization for the quality trials below.

Basically, if you're looking for top quality at aggressive bit rates, you shouldn't use Apple's codec. On the other hand, different implementations of the open-source x264 codec bear different compatibility risks, which I think are manageable if you pay attention and test to avoid these problems. For example, when I first tested x264 back in March ("If you're encoding in QuickTime/Compressor, you gotta check out x264," **bit.ly/checkoutx264**), several posters left comments about compatibility issues they experienced with x264.

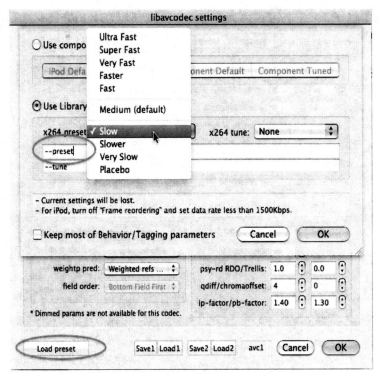

Figure 8-2. Settings used for the x264Encoder.

However, the x264 codec is (reportedly) used by YouTube and other UGC sites, so it must be very compatible when properly configured. If you try x264—and you should if you're currently using Compressor—you should test your encoded streams on all downstream systems that your video will be viewed upon, including iDevices, Apple TV, QuickTime Broadcaster and others. Anyway, except as otherwise noted, all comparison shots that follow use x264, rather than the Apple codec, since Apple lags quality-wise in every challenging test file.

Sorenson Squeeze 7.0.0.126

There were several items to consider with Squeeze on both the Mac and Windows platforms. First, which version to test? I started out testing version 6.5, then got the beta of version 7, then the shipping version of Squeeze 7. There are two key H.264-related features differentiating versions 6.5 and 7. First, with 6.5, Squeeze can only encode one H.264 file at a time; with version 7, it can encode multiple H.264 files in parallel. (Note: You could encode VP6 files in parallel in Squeeze 6.5, but not H.264.) Though the single file encoding times were very similar between the programs, parallel encoding swung it for me, so I tested with version 7.

The other major difference between the two versions is that Squeeze 7 offers a new GPU-accelerated version of the MainConcept H.264 encoder in addition to the normal, CPU-only version that's identical to version 6.5. This is the CUDA version you see in Figure 8-3.

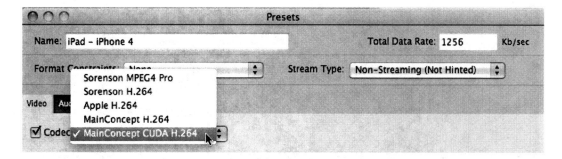

Figure 8-3. Squeeze 7's new CUDA version of the MainConcept codec is much faster than the non-accelerated versions in Squeeze 6.5.

The CUDA accelerated version is much faster than even two-pass encoding (rendering in 3:20 min:sec rather than 7:28), and the quality was generally very good. However, there were also several regions where CUDA-accelerated encoding was clearly deficient compared with the software-only encoding available in versions 6.5 and 7. You can see this in Figure 8-4.

Figure 8-4. Sometimes, faster isn't better.

Sorenson is aware of the problem and will resolve it with a free upgrade that should be available by mid-2011. In the meantime, since I don't recommend using CUDA-accelerated encoding until this issue is resolved, I didn't use this option for either the performance or encoding trials.

The next issue with Squeeze was whether to use two-pass VBR or multi-pass encoding mode, which requires five passes through the data and takes a lot longer. For example, on the Mac, multi-pass took 14:20 to produce the SD test file, compared with 7:28 for VBR (both min:sec). The obvious question is whether the additional time is worth it.

In my comparison tests, it didn't seem to be. Figure 8-5 shows a bellwether frame from the SD test clip, and there's no visible difference between the results from the two different techniques. Accordingly, I produced all Squeeze test files on the Mac and in Windows using two-pass VBR rather than multi-pass encoding.

Figure 8-5. I saw very little difference between two-pass VBR and multi-pass, so I used the former.

In the SD trials, I used the parameters shown in Figure 8-6. Other than Data Rate and Frame Size, these are the same parameters that I used for the HD encoding.

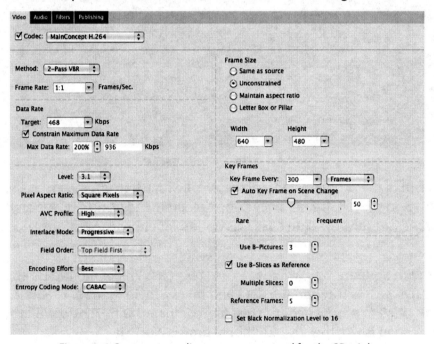

Figure 8-6. Squeeze encoding parameters used for the SD trials.

Though I typically don't use B-Slices as a reference (B-Slices are the same as B-frames; I'm just using the term from the Squeeze UI) because some experts claim that it can cause playback instability, I enabled it for these trials to maximize output quality. Otherwise, to recap, I encoded in Squeeze 7 using the CPU-only, with two-pass VBR for all encodings.

Telestream Episode Pro 6.1.1.2

Telestream offers three versions of its desktop encoding program: Episode ($495), Episode Pro ($995) and Episode Engine ($3,995). The latter two can import more high-end formats and encode in parallel, though Episode Pro is limited to two simultaneous encodes, while Episode Engine can simultaneously encode as many files as the number of cores in your computer. I used Episode Pro for these tests, and Figure 8-7 shows the settings used.

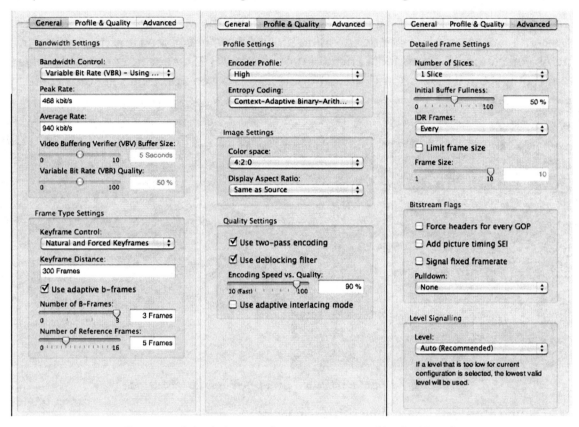

Figure 8-7. Episode Pro encoding parameters used for the SD trials.

Note that in the move from version 6.0 to 6.1.1, which Telestream released in early April 2011, Telestream removed the constrained VBR option from the H.264 encoding interface. When I asked why, the product manager related that the MainConcept codec didn't respond to the constraint-related commands, so the company just removed the controls. Accordingly, when you encode in Episode Pro for any scenario where you'd typically use constrained VBR, like mobile or adaptive streaming, use CBR.

Adobe Media Encoder Version 5.0.1.0

I did most of my testing with version CS5, but tested late beta of CS5.5 to make sure that there were no dramatic quality, performance or H.264 feature-related changes (there weren't). Just to mix things up, Figure 8-8 shows the settings used for the HD encodes in Adobe Media Encoder. Note that the controls don't live like this in the program—I lopped off the bottom and placed it alongside the top to save space.

When I encode with Adobe Media Encoder, I don't enable the Maximum Render Quality check box because it dramatically extends encoding time and I've never seen any quality improvement. As you can see, Adobe Media Encoder provides few H.264-specific controls—profile and level only—but it makes some interesting decisions based upon the selected profile.

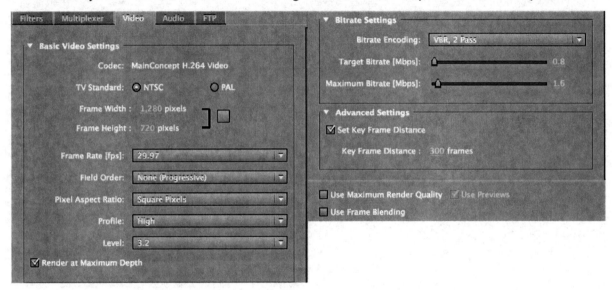

Figure 8-8. Adobe Media Encoder parameters used for the HD trials.

Here are some observations from files that I analyzed:

• When you choose the Baseline profile, the file has no B-frames and uses CAVLC entropy encoding.

• When you choose the Main profile, the encoder deploys CABAC entropy encoding and a B-frame interval of 3 for an IBBBPBBBPBBBP ... cadence.

- When you select the High profile, Adobe Media Encoder deploys CABAC for entropy encoding, with a B-frame interval of two, for an IBBPBBPBB ... and so on cadence. Semaphore also reported that there were three reference frames.

Quality Results: Mac Encoders

Before discussing qualitative results, let's take a quick look at the tests that I used: two for quality, four for performance. For the record, I performed all Mac encoding trials on my 8-core (2x2.93 GHz quad-core Intel Xeons) Mac Pro with 12 GB of RAM, running Mac OS X 10.6.5. The graphics card was an NVIDIA Quadro FX 4800, though I don't think it accelerated any encodings. Here are relevant details about each encoder:

- **SD test file**: The SD test file is 5:53 (min:sec) long, composed of 42 DV source sequences that were prescaled and pre-deinterlaced into a 640x480@29.97 progressive file. All I did in the encoding tool was encode; no scaling or deinterlacing. Target encoding parameters were 468 kbps video/32 kbps mono audio, which computes to a bits-per-pixel value of around .05, which is aggressive. In comparison, about the lowest that I've seen in any deployed SD video is around .05, with most in the .1 and higher range. After encoding the file, I checked to make sure that it was within 5% of the target data rate, plus or minus.

I used the highest profile available, usually the High profile, and used two-pass constrained VBR with maximum data rate set to 2x the target, or around 940 kbps. I maximized all quality settings in all encoders (unless otherwise discussed), deploying CABAC when available, with B-frame settings at an interval of three, and five reference frames. I enabled reference and pyramid B-frames when available, but I always encoded using a single slice. Except as mentioned, I set the key frame interval at 300 with scene change detection enabled.

- **HD test file:** The HD test file is 1:32 long, composed of seven HDV sequences that were pre-scaled and pre-deinterlaced into a 1280x720@29.97 progressive file. All I did in the encoding tool was encode; no scaling or deinterlacing. Target encoding parameters were 800 kbps video/128 kbps stereo audio, for a video bits-per-pixel value of around .03, which is very aggressive. In comparison, about the lowest that I've seen in any deployed HD video is around .03, with most in the .1 and higher range. After encoding the file, I checked to make sure that it was within 5% of the target data rate, plus or minus.

As with the SD file, I used the highest profile available, usually the High profile, and used two-pass constrained VBR with maximum data rate set to 2x the target, or around 1600 kbps. I maximized all quality settings in all encoders (unless otherwise discussed), deploying CABAC when available, with B-frame settings at an interval of three, with five reference frames. I enabled reference and pyramid B-frames when available, but I always encoded using a single slice. Except as mentioned, I set the key frame interval at 300 with scene change detection enabled.

In terms of performance, I timed how long it took for each program to encode the SD and HD files. In addition, I ran these tests:

- **One-to-many trial:** In the one-to-many trial, I used the HD test file and produced eight files: a 5 Mbps file for YouTube, three files for iDevices (320x180, 640x360 and 720p) and four files for adaptive streaming, all at 640x360 resolution at 500, 800, 1200 and 1500 kbps. I won't bore you with the individual encoding parameters for each file, but they were applied as consistently as possible. I did not judge the quality of these encoded files; I just timed the performance.

- **Many-to-one trial:** In the many-to-one trial, I used eight 1-minute DV source clips encoded using the preset used for the SD trials. In the interest of time, I did not deinterlace these files (too many settings to change), but I did scale all files from 720x480 to 640x480. With this test, if the encoder produced serially rather than in parallel and could load multiple instances, I encoded twice: once encoding all eight files in a single instance, once loading eight instances, and encoding a file in each. If all this instance stuff sounds like gobbledygook to you, check out "Accelerating Encoding on Multiple-Core Workstations" (bit.ly/multicore_encode), or Chapter 12. I did not judge the quality of these encoded files.

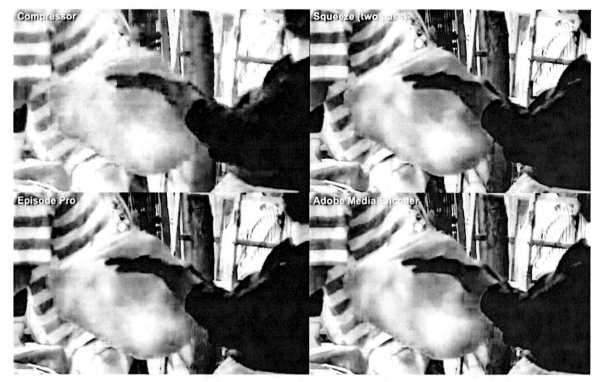

Figure 8-9. These comparative frames show that the Apple codec retains much less quality than the other encoders.

To assess quality, I loaded the test files that I created into a Premiere Pro sequence so I could compare the frame quality side by side. I work with four files at a time because that's most convenient, and provides a quadrant like that shown in Figure 8-9. I compared the frames at their original resolution, though the image in Figure 8-9 was edited down for more convenient presentation in this book. As you can see, the Compressor frame, using the Apple codec, is clearly degraded compared with the others.

After assessing single frame quality, I rendered the Premiere Pro sequences using a lossless codec (or ProRes on the Mac) and played the timeline to assess the presence or absence of motion artifacts in the respective clips.

Figure 8-10. These comparative frames show very similar quality among all encoders.

With HD files, which I encoded at 1280x720, I used the same 1280x720 sequence setting, since a 100% 4x4 presentation would require an unwieldy 2560x1440 sequence. Then I cropped each video within the sequence to show the most important content in the smaller window. That's what you see in Figure 8-10, which shows that all encoders produced very similar quality, with the x264Encoder perhaps a hair behind.

All qualitative comparison systems are imperfect, but this technique allows me to compare frame-by-frame quality as well as motion playback quality with two test clips that I've used for a combined eight years or so. I'm also in frequent contact with most encoding vendors (excluding Apple) to discuss and verify these findings, which gives me a level of comfort that I'm not totally off-base.

With this as prologue, on to Table 8-2. As mentioned, avoid the Apple codec if possible, substituting in the x264Encoder if you want to preserve the Compressor workflow but verifying compatibility with downstream systems and players. In contrast, Adobe Media Encoder does a great quality-wise for an essentially free encoding tool, but showed just a touch less quality in some HD frames. Nothing any viewer would notice, but definitely there.

	Adobe Media Encoder	Apple Compressor	Apple Compressor with x264Encoder	Sorenson Squeeze	Telestream Episode Pro 6
SD frame quality	Excellent	Fair	Very Good	Excellent	Excellent
SD motion quality	Excellent	Good	Very Good	Excellent	Excellent
HD frame quality	Very Good	Poor	Very Good	Excellent	Excellent
HD motion quality	Excellent	Fair	Very Good	Excellent	Excellent
Bottom line	Just a hair behind the leaders in HD quality	Avoid if possible	Very, very good, but verify compatibility	The standard by which others are judged	Version 6.1.1 brings parity with Squeeze

Table 8-2. Encoding quality of Mac-based encoders.

Episode Pro version 6 is the first generation of Episode with the MainConcept codec, which is a significant improvement over the Dicas codec used previously. While there were some rough edges in the 6.0 release, they were all resolved in version 6.1.1, which brought Episode Pro to parity with Squeeze quality-wise. Otherwise, the x264Encoder was just a hair behind Squeeze and Episode Pro.

As mentioned, I tested a late beta version of the Adobe Creative Suite 5.5 on both platforms to make sure that the encoding results and features didn't change, but I didn't do the same for Final Cut Pro X, which Apple announced at NAB in April 2011. This book went to the publisher days after that announcement, and Apple hasn't even specified whether Compressor will be upgraded with FCP. I'll figure this out in due time and will post my findings at both StreamingMedia.com, and at StreamingLearningCenter.com.

Now let's look at encoding speed.

Encoding Speed: Mac Encoders

Quality is one thing; encoding speed another. Table 8-3 tells the tale.

	Adobe Media Encoder	Apple Compressor	Apple Compressor with x264Encoder	Sorenson Squeeze	Telestream Episode Pro 6
Encoding:Serial or Parallel	Serial	Parallel (with Qmaster)	Parallel (with Qmaster)	Parallel	Parallel (2 files max)
Single SD file to 500 kbps	2:47	8:21	5:30	7:28	3:17
Single HD file to 800 kbps	2:10	4:30	3:00	3:47	2:14
Single HD test file to 8 files	15:55	23:36	34:30	24:30	15:17
8 DV files to 500 kbps	2:48	3:29	3:02	5:20	2:27

Table 8-3. Mac performance results (min:sec).

The results tend to speak for themselves, with no breakout times in any direction. Interestingly, though Adobe Media Encoder is the only program that encodes exclusively in serial mode, it does so extremely quickly, resulting in very competitive times. As between the Apple H.264 codec and the x264Encoder, it's nice that the quality upgrade comes with faster performance in three out of the four tests. Episode Pro is faster than Squeeze—significantly so in some tests—which should be considered by high-volume producers, particularly now that Episode Pro offers similar quality.

Encoding Features: Mac Encoders

Let's consider other features available in the respective encoders, starting with H.264 configuration options. As you can see in Table 8-4, the only encoder that provides extensive configurability is the x264Encoder. I've included two of the four configuration screens in Figure 8-11.

	Adobe Media Encoder	Apple Compressor	Apple Compressor with x264Encoder	Sorenson Squeeze	Telestream Episode Pro 6
Baseline/Main/High profile selection?	Yes	Baseline/Main	Yes	Yes	Yes
CABAC entropy encoding	Automatic	No	Yes	Yes	Yes
B-frame/reference frame controls	Automatic	Automatic	Yes	Yes	Yes
Other H.264 controls	None	None	Very extensive	Minimal	Minimal
Adaptive streaming presets	Yes	No	No	Yes	No

Table 8-4. H.264 encoding controls for Mac-based encoders.

Who needs this level of configurability? Several groups. First, if you're repeatedly encoding very idiosyncratic footage, like a soccer or baseball game from a set camera, you may want

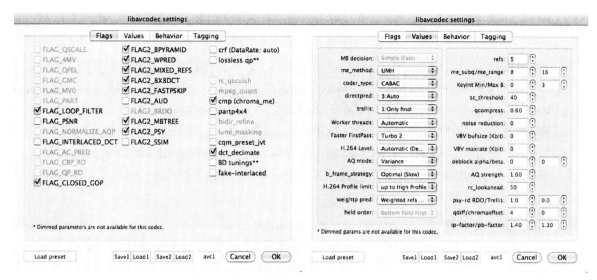

Figure 8-11. Advanced controls of the x264Encoder.

to adjust these options to optimize quality for that footage. If you're encoding massive quantities of relatively generic footage, as in a UGC site, you may want to tweak controls like these to achieve the desired blend of quality and data rate.

Fortunately, the developer of the x264Encoder, Takashi Mochizuki, does a nice job detailing the available configuration options in the documentation that comes with the codec. Still, even for an advanced user, it's going to take a lot of time and effort to derive configuration options that visibly outperform the presets that Mochizuki provides.

Moving to the last item in Table 8-4, with version 7, Sorenson Squeeze introduced adaptive streaming presets that let you encode a single file to multiple files, though only for streaming to iDevices. Sorenson does plan to release additional adaptive streaming presets down the road. Unlike Sorenson's adaptive streaming presets, Adobe's presets don't encode a single file to multiple outputs; you have to load the source file multiple times and apply each profile individually, which is less convenient. You can read more about the Adobe adaptive streaming presets at bit.ly/adaptivepresets.

Automation and Input/Output

Now let's look at the automation and input/output features available with the respective programs. Here's a brief explanation of the various features:

- **Droplets**: encoding presets encapsulated in a desktop icon—users drag a file onto the icon to start encoding. A nice ease-of-use feature.

- **Watch folders**: folders associated with encoding presets that the encoding tool "watches." Once files are dropped in, the encoder renders them as designated in the preset. A great sharing feature if users on a network can drop files into the shared watch folder.

- **Access from Compressor**: the ability to access the encoder's features from within Compressor, typically either as a codec (x264Encoder) or QuickTime Export Component. This is great if your current workflow involves Compressor and you don't want to change the workflow.

- **Access from Adobe Media Encoder**: the ability to access the encoder's features from within Adobe Media Encoder. Again, this is very convenient if you prefer a workflow that involves Adobe Media Encoder.

- **Retrieve via FTP**: the ability to retrieve a file from a remote FTP site and start encoding (like a watch folder, but via FTP).

- **Cluster encoding**: the ability to share encoding tasks with other computers on a network. This is great for high-volume shops with multiple encoding stations.

- **Deliver via FTP**: the ability to deliver an encoded file via FTP.

- **Upload to YouTube**: the ability to log in to YouTube, upload and input user metadata.

- **Email/text notification**: the ability to trigger messages upon completion or failure.

	Adobe Media Encoder	Apple Compressor	Apple Compressor with x264Encoder	Sorenson Squeeze	Telestream Episode Pro 6
Droplets	No	Yes	Yes	No	No
Watch folders	Yes	Via Applescript	Via Applescript	Yes	Yes
Access from Compressor	No	Yes	Yes	No	No
Access from Adobe Media Encoder	Yes	No	No	Yes	No
Retrieve via FTP	No	Via Applescript	Via Applescript	No	Yes
Cluster encoding	No	Yes	Yes	No	Yes
Deliver via FTP	No	Via Applescript	Via Applescript	Yes	No
Upload to YouTube	No	Yes	Yes	Yes	Yes
Email/text notification	No	Yes	Yes	Yes	No

Table 8-5. Automation and input/output features of Mac encoding tools.

The answers in the Table 8-5 are self-explanatory, save the "Via Applescript" in the Compressor columns. If you're a programmer (or one of your BFFs is), this means that you can probably create a script that will accomplish the specified task. If you're a mere mortal and all of your buddies (and co-workers) are as well, this means that you can't accomplish the task using normal program controls.

Other Features

Table 8-6 captures features that may be important to buyers of a batch-encoding program. Here's a brief explanation of these features:

- *Intro/outro*: the ability to add videos to the start and end of the clip, primarily used for branding and/or advertising. Obviously, Adobe Media Encoder and Apple Compressor can perform this task in their respective editors, though it's more convenient to perform this during encoding.

- *Watermark/logo insertion*: the ability to add a logo overlay over the clip, primarily for branding or content protection. "In editor" in the Adobe Media Encoder column means that you can do this in Premiere Pro, but not in Adobe Media Encoder (Compressor has a watermark function and doesn't need Final Cut Pro).

- *Telecine removal*: the ability to remove extra fields that were input into film-based content for DVD or VHS production. Important to those encoding film-based content that had been converted to NTSC.

- *Add metadata*: the ability to add metadata to the file before encoding.

Other Features:	Adobe Media Encoder	Apple Compressor	Apple Compressor with x264Encoder	Sorenson Squeeze	Telestream Episode Pro 6
Intro/outro	In editor	In editor	In editor	No	Yes
Watermark/logo insertion	In editor	Yes	Yes	Yes	Yes
Telecine removal	No	No	No	Yes	Yes
Add metadata	Yes	No	No	No	Yes

Table 8-6. Other features of the Mac-based encoding tools.

That's it for the Mac-based software tools; let's look at the Windows based tools.

Windows Encoders

I performed all Windows encoding trials on my 12-core (2x3.33 GHz six-core Intel Xeons) HP Z800 with 24 GB of RAM, running 64-bit Windows 7 Professional, with an NVIDIA Quadro FX 4800 graphics cards. Here are relevant details about each encoder.

	Adobe Media Encoder	Microsoft Expression Encoder	Sorenson Squeeze	Telestream Episode Pro 6
Platforms	Mac/Windows	Windows	Mac/Windows	Mac/Windows
Price as tested	Bundle-only	$199.95	$799	$995
H.264 codec	MainConcept	MainConcept	MainConcept	MainConcept
Other streaming codecs supported	VP6, WMV	WMV	VP6, WMV	VP6, WMV

Table 8-7. Meet the Windows encoding tools.

Three of the Mac contenders also play on Windows: Adobe Media Encoder, Sorenson Squeeze and Telestream Episode Pro. I used the same procedures for Windows as I did on the Mac, so Windows users can read about these in the preceding section. Apple Compressor is Mac-only, of course, so I've substituted Microsoft's Expression Encoder 4, which retails for $199.95 and runs only on Windows.

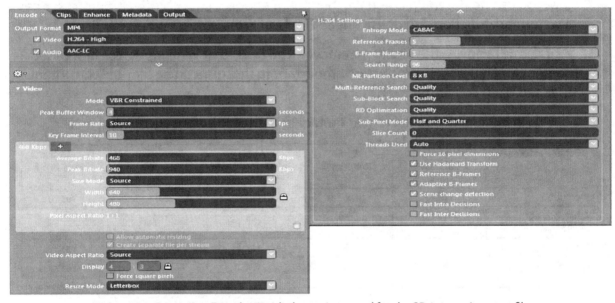

Figure 8-12. Expression Encoder 4 with the settings used for the SD comparison test file.

With version 4, Microsoft began using the MainConcept H.264 codec rather than their own, which allowed Microsoft to enable the High profile and GPU encoding—though this is the same GPU encoding that we saw in Squeeze 7, and Expression Encoder produced the same ugly frames as Squeeze, so I tested Expression Encoder with GPU encoding disabled. Expression Encoder encodes serially, but you can run multiple instances, which I did for the final performance test. Figure 8-12 shows the encoding settings used for the SD encoding trials.

Quality Results: Windows Encoders

Table 8-6 shows the results of my quality tests. In these trials, Expression Encoder, Episode Pro and Squeeze were neck and neck at the top of the leader board, with Adobe Media Encoder just a hair behind in HD quality (Figure 8-13) but virtually identical in SD quality (8-14).

	Adobe Media Encoder	Microsoft Expression Encoder	Sorenson Squeeze	Telestream Episode 6
SD frame quality	Excellent	Excellent	Excellent	Excellent
SD motion quality	Excellent	Excellent	Excellent	Excellent
HD frame quality	Very Good	Excellent	Excellent	Excellent
HD motion quality	Excellent	Excellent	Excellent	Excellent
Bottom line	Just a touch behind in some HD frames	A solid choice, particularly if you like to tinker	The standard by which others are judged	Version 6.1.1 brings parity with Squeeze

Table 8-8. Encoding quality of Windows-based encoders.

Figure 8-13. The HD test file showed Adobe Media Encoder just a hair behind the others.

I hate to keep showing the same frame, but Figure 8-14 combines high-motion with extensive detail, and is a bellwether frame. Here you can see the excellent quality produced by all four MainConcept-based encoding tools.

Figure 8-14. The SD test clip showed all encoders neck and neck.

Now let's look at encoding speed.

Encoding Speed: Windows Encoders

Table 8-9 contains the performance results for the Windows encoders. While both Adobe Media Encoder and Expression Encoder produced their files serially, operation was fairly snappy for both programs. I did shave 9 seconds with Expression Encoder by loading eight instances and encoding separately, which obviously wasn't worthwhile since the extra set up cost me about five minutes. Still, if you have several large files to encode, the time savings might be worthwhile. You can't load multiple instances of Adobe Media Encoder, but multiple-file encoding speed was super nonetheless.

	Adobe Media Encoder	Microsoft Expression Encoder	Sorenson Squeeze	Telestream Episode Pro
Encoding: serial or parallel	Serial	Serial	Parallel	Parallel (2 files max)
Single SD file to 500 kbps	1:30	2:34	3:46	1:21
Single HD file to 800 kbps	1:33	1:19	2:42	3:26
Single HD test file to 8 files	12:16	7:40	16:59	10:53
8 DV files to 500 kbps	2:16	6:04	3:23	2:34
8 DV files - multiple instances	NA	5:55	NA	NA

Table 8-9. Windows performance results (min:sec).

Otherwise, the results are fairly consistent between the encoders, without real outliers to dramatically change your opinion about any of the products.

Encoding Features: Windows Encoders

Now lets look at H.264 encoding options, as shown in Table 8-10. As it turns out, Expression Encoder offers the most extensive H.264 encoding controls, plus the most effective presentation of adaptive streaming presets, though they can only be used for streaming to a Windows Media or Silverlight player, not Flash. You can see the adaptive streaming options in Figure 8-15, each with its own little tab.

H.264 Encoding Features	Adobe Media Encoder	Microsoft Expression Encoder	Sorenson Squeeze	Telestream Episode Pro
Baseline/Main/High profile selection?	Yes	Yes	Yes	Yes
CABAC entropy encoding	Automatic	Yes	Yes	Yes
B-frame/reference frame controls	Automatic	Yes	Yes	Yes
Other H.264 controls	None	Very extensive	Minimal	Minimal
Adaptive streaming presets	Yes	Yes	Yes	No

Table 8-10. H.264 encoding controls for Windows-based encoders.

Sorenson Squeeze introduced adaptive streaming presets with version 7, though only for streaming to iDevices, and they're configured like Expression Encoder, where one preset produces multiple output files. Unlike Squeeze and Expression Encoder, Adobe Media Encoder doesn't encapsulate multiple encoding profiles in a single preset; instead it provides multiple presets that you can apply to a single file. You can read more about the Adobe adaptive streaming presets at **bit.ly/adaptivepresets**.

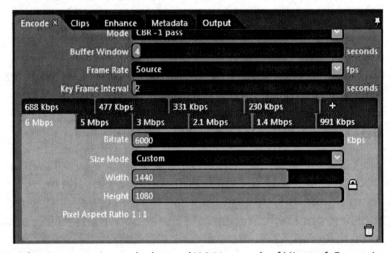

Figure 8-15. Adaptive streaming and advanced H.264 controls of Microsoft Expression Encoder 4.

Automation and Input/Output

Now let's look at the automation and input/output features available with the respective programs. Here's a brief explanation of the various features:

- **Watch folders**: folders associated with encoding presets that the encoding tool "watches." Once files are dropped in, the encoder renders them as designated in the preset. A great sharing feature if users on a network can drop files into the shared watch folder.

- **Access from Adobe Media Encoder**: the ability to access the encoder's features from within Adobe Media Encoder. This is very convenient if you prefer a workflow that involves Adobe Media Encoder.

- **Retrieve via FTP**: the ability to retrieve a file from a remote FTP site and start encoding (like a watch folder, but via FTP).

- **Cluster encoding**: the ability to share encoding tasks with other computers on a network. This is great for high-volume shops with multiple encoding stations.

- **Deliver via FTP**: the ability to deliver an encoded file via FTP.

- **Upload to YouTube**: the ability to log in to YouTube, upload and input user metadata.

- **Email/text notification**: the ability to trigger messages upon completion or failure.

	Adobe Media Encoder	Microsoft Expression Encoder	Sorenson Squeeze	Telestream Episode Pro
Watch folders	Yes	No	Yes	Yes
Access from Adobe Media Encoder	Yes	No	Yes	No
Retrieve via FTP	No	No	No	Yes
Cluster encoding	No	No	No	Yes
Deliver via FTP	Yes	No	Yes	Yes
Upload to YouTube	No	No	Yes	Yes
Email/text notification	No	No	Yes	No

Table 8-11. Automation and input/output features of the Windows-based encoding tools.

Other Features

The next table captures other features that may be important to buyers of a batch-encoding program. Here's a brief explanation:

- **Intro/outro**: the ability to add videos to the start and end of the clip, primarily used for branding and/or advertising. Obviously, Adobe Media Encoder can perform this task in Premiere Pro, though it's more convenient to perform this during encoding.

- **Watermark/logo insertion**: the ability to add a logo overlay over the clip, primarily for branding or content protection. "In editor" in the Adobe Media Encoder column means that you can do this in Premiere Pro, but not in Adobe Media Encoder.

- **Telecine removal**: the ability to remove extra fields that were input into film-based content for DVD or VHS production. Important to those encoding film-based content that had been converted to NTSC.

- **Add metadata**: the ability to add metadata to the file before encoding.

	Adobe Media Encoder	Microsoft Expression Encoder	Sorenson Squeeze	Telestream Episode 6
Intro/outro	In editor	Yes	No	Yes
Watermark/logo insertion	In editor	Yes	Yes	Yes
Telecine removal	No	Yes	Yes	Yes
Add metadata	Yes	Yes	No	Yes

Table 8-12. Other features of the Windows-based encoding tools.

That's it for Windows-based encoding tools.

Conclusion

Hopefully, this chapter provides some useful information about how to choose and use an H.264 encoding tool. Next chapter, I'll detail how to produce video using other codecs, including VP6 and WebM.

Chapter 9: Producing WebM, VP6 and WMV Files

While H.264 is the most important codec these days, it's not the only codec, so in this chapter I'll detail how to produce video with other codecs, including WebM, VP6, Windows Media and Ogg Theora (briefly).

Let's start with the "it" codec in waiting, WebM.

Producing WebM

I introduced you to WebM back in Chapter 3 in the section titled Google's WebM. Briefly, WebM is an open-source video format that uses the VP8 video codec, the Vorbis audio codec and a file structure based upon the open-source Matroska container format. In terms of video quality, WebM is very close to H.264, though it takes longer to encode than H.264 and requires more CPU horsepower to decode than H.264 on computers with H.264 GPU acceleration.

If you're producing video for the HTML video tag, H.264 and WebM are the two primary alternatives. You've learned all about H.264; now we'll tackle WebM.

As an overview, there are multiple options for encoding WebM. You can read about the free tools at www.webmproject.org. Here I'll describe three encoding alternatives: one free, and two that you'll have to pay for.

By way of background, I tested five WebM encoding tools for an article that I wrote for *StreamingMedia* Magazine, which you can read at bit.ly/webmencoderreview. Of the five, two tools—Miro Video Converter and Wildform Flix WebM—were totally not production-worthy.

Of the three other encoders, Firefogg was close, but it had serious issues delivering files that matched the input parameters. Telestream Episode Pro produced very good quality, but the files "broke" when played back in Google Chrome. Sorenson Squeeze was the best option, with very good quality, very good access to WebM's extensive encoding options and no serious defects.

Since the article is available online, what I'll do here is update those results and provide additional details regarding how to produce WebM footage in all tools. I'll also relate some performance data and let you know if the issues that I raised in the review for Firefogg and Episode have been resolved (the quick answer, unfortunately, is no).

Firefogg

Firefogg was the only free alternative I tested that was even close to viable, though the command line tools available at www.webmproject.org undoubtedly work. Firefogg is a Firefox plug-in that you can install by navigating over to www.firefogg.org in Firefox and following the installation instructions on the site.

Once the plug-in is installed, to encode a file, you click Make web video, then Select file, which opens the standard Select file screen on either Mac or Windows. In the Custom Settings drop-down list (Figure 9-1), choose one of the WebM alternatives: the first, WebM web video VP8 (600 kbit/s and 480px maximum width), for SD video; the other for HD.

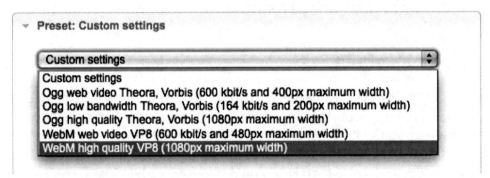

Figure 9-1. Choosing the WebM codec.

Then, customize the settings as desired. Set the Encoding range if you don't want to encode the entire file, and then click the Basic quality and resolution control (Figure 9-2). Don't bother with Video quality if you plan to insert a data rate in the Advanced Video Encoding controls, since that takes precedence.

Though the Audio quality informational dialog gives you the same message about setting the audio data rate, when you click over to the Advanced Audio Encoding Controls, the only option is No audio—there is no data rate control. Or at least it was this way for me during both sets of tests; perhaps it's been updated, so you should have a look. If not, you have to adjust the quality level to match the desired audio output, which will take some trial and error.

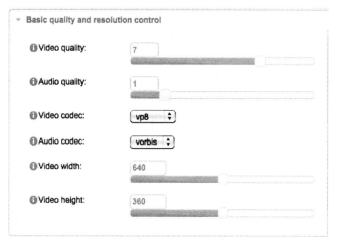

Figure 9-2. Setting basic quality and resolution controls parameters.

Then click the Advanced Video Encoding Controls and set your configuration options there (Figure 9-3). Shown in the figure are the options for my standard SD test file.

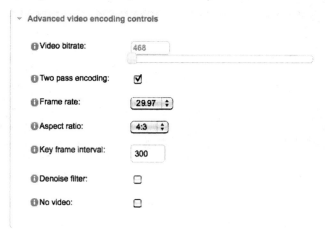

Figure 9-3. Setting advanced video encoding controls.

When you're finished setting the desired options, click Encode to File, and Firefogg will open a Save Video As dialog that you can use to name and choose the location for your file. Once you click Save, the encoding starts.

Alas, my retesting was in vain, since Firefogg's output shared the same flaws found in my *StreamingMedia* review; specifically, though I asked for a 640x480 file, Firefogg produced the file at 1080x810 and at 3401 kbps rather than 468. I didn't retest the HD video file, which Firefogg produced very well during my earlier tests.

If you're looking for a production-quality tool, Firefogg ain't it, but if Firefogg resolves its resolution issues and lets you dial in an audio data rate, it could be a very valuable tool. If you're looking for a free, simple-to-use tool for experimenting with WebM, give it a shot.

Telestream Episode Pro

Telestream Episode Pro has an exceptionally simple WebM encoding interface—basically, no WebM configuration options at all; you choose the codec, and that's it. Though output quality was very good—Telestream made some great choices for default WebM encoding options—in version 6.0, the videos became distorted when played back in the Google Chrome browser, as you can see in Figure 9-4.

Figure 9-4. Telestream resolved this playback problem in Chrome in Version 6.1.1.

Fortunately, Telestream fixed this problem in Version 6.1.1, so if you're producing WebM videos with Episode Pro and see the corruption shown in Figure 9-4, make sure that you're upgraded to the latest version. Episode was also the fastest encoder, producing my 5:54 (min:sec) SD test file in 12:35, as compared with 14:45 for Squeeze and 17:50 for Firefogg (all tests on my 2.93 GHz 8-core Mac Pro).

If you like to tinker with your encoding settings, Episode Pro is not your tool. On the other hand, if your priorities are high quality and fast encoding, give it a shot.

Sorenson Squeeze

Sorenson Squeeze produced similar WebM quality to Episode, and though it was a little slower, it offers the most extensive WebM configuration options of any non-command line WebM encoding tool that I've seen. Let's spend some time exploring the encoding parameters that I used, which were guided by recommendations from the WebM documentation (www. webmproject.org/tools/encoder-parameters).

I was testing while Sorenson was finalizing its presets, and we worked together to develop the optimal values, but I'm not sure that my recommendations made their way into each and every preset. So I'll let you know which values the codec designers recommend, and you can make sure the parameters in the Squeeze templates are correct. Let's start with the options shown in Figure 9-5.

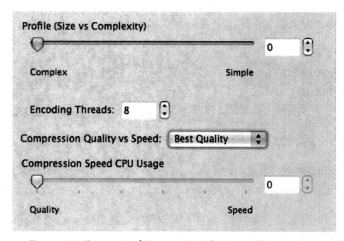

Figure 9-5. Screen 1 of Squeeze's WebM encoding options.

- **Profile.** Recommended value: 0. Here's the relevant blurb from the WebM docs:

  > For non-zero values the encoder increasingly optimizes for reduced complexity playback on low-powered devices at the expense of encode quality. For example using 1 tells the encoder to use only bi-linear sub pixel filtering and a simplified loop filter. In general most users should set a value of 0 or ignore this parameter unless they are encoding high-res content and require playback on very low-power devices.

- **Encoding Threads.** Recommended value: number of real cores on your system, minus 1. I should have used 7 on the 8-core Power Mac, but I wanted to maximize encoding speed. The concern with multiple threads, of course is a diminution in quality. However, in testing reported at bit.ly/webmcloserlook, I tested with eight cores and with one core, and saw no quality difference. In the meantime, on my HP Z800 with two 6-core, 3.33 GHz Xeon processors, encoding with one thread took 41:40, while 12 threads took 13:16 (both times min:sec). So on a multi-core system, you definitely want to go with the recommendation. I think that Sorenson's default value for Encoding Threads will be 30, which should be just as fast and shouldn't degrade quality, but I'll go with number of cores minus one in my encodes.

- **Compression Quality vs. Speed.** Recommended value: Good quality, 0 speed. I was testing for optimal quality, so I used Best Quality. Google's docs disagree, stating: "In general this is not a recommended setting unless you have a lot of time on your hands." If you do select Good Quality, you can drag the Compression Speed CPU Usage to five different values that trade off quality for encoding speed. The docs say "setting -- good quality and --cpu-used=0 will give quality that is usually very close to and even sometimes better than that obtained with --best but the encoder will typically run about twice as fast.

  In a production environment, I would go with Good Quality, 0 CPU, which I think will also be Squeeze's default setting.

Figure 9-6 shows the next set of options. The first setting, VBR Variability, controls the amount of variance in the stream: it's a Sorenson option, not a WebM option. The default value for VBR files is 70, and I typically use that. The control is grayed for CBR files. The next two settings you learned back in <u>Chapter 2</u>, and 50 and 200 are my recommended options.

The next few options are WebM-specific and Figure 9-6 shows my recommendations.

- **Data Rate Undershoot.** Recommended value: 100 for two-pass encodes and when producing in VBR mode. If you're producing in CBR, you're on your own, because there's no guidance given.

- **Minimum Quality.** Recommended value: 0-4.

- **Maximum Quality.** Recommended value: 50-63.

- **Automatically use Alternate Reference Frames.** Recommended value: enabled. It's disabled by default. I enabled for my tests, and recommend that you do as well. Here's what the Google documents say: "Use of --auto-alt-ref can substantially improve quality in many situations (though there are still a few where it may hurt)."

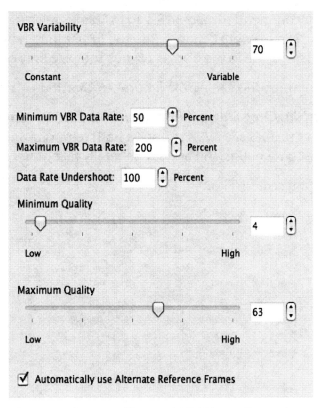

Figure 9-6. Screen 2 of Squeeze's WebM encoding options.

On the home stretch, Figure 9-7 contains the penultimate controls.

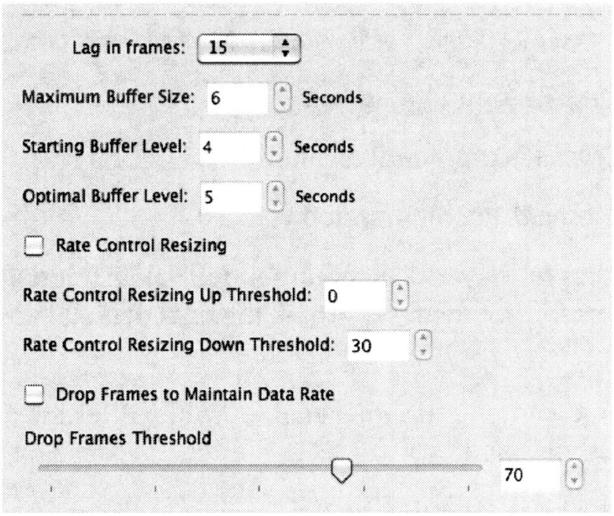

Figure 9-7. Screen 3 of Squeeze's WebM encoding options; almost home.

- **Lag in Frames.** Recommended value: 15. Lag in Frames relates to Alternate Reference Frames, detailing the maximum distance into the future that the encoder can look for the frame. The default value in the docs is 16, but Squeeze only goes up to 15.

- **Buffer Size.** Recommended values: Maximum—6, Starting—4, Optimal—5.

- **Rate Control Resizing/Drop Frames to Maintain Data Rate.** Recommended value: disabled. Here's what the docs say: "VP8 supports both temporal and spatial resampling. These are specialist parameters and are not generally recommended."

Next up: the final encoding screen (Figure 9-8).

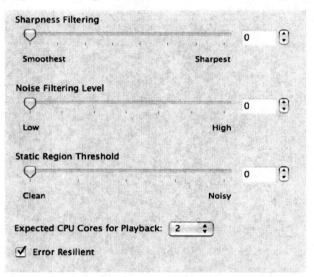

Figure 9-8. Screen 4 of Squeeze's WebM encoding options.

- **Sharpness Filtering.** Recommended value: 0.

- **Noise Filtering Level.** Recommended value: 0.

- **Static Region Threshold.** Recommended value: 0.

- **Expected CPU Cores for Playback.** Recommended value: not defined in the Google documentation. I used 1 for my tests. I think the final default in Squeeze will be 2; if so, go with that.

- **Error Resilient.** Recommended value: enabled when Alternate Reference Frames are enabled; otherwise, disabled.

That's it for WebM; let's turn our attention to VP6.

Producing VP6

Though overshadowed by H.264 in the news, On2's VP6 is still the most widely used streaming codec in the world today. In this section, you'll learn what you need to know to produce video using the On2 VP6 codec.

On2, VP6 and VP8

By way of background, the first big-time Flash codec was Sorenson Spark, which YouTube still uses for its lowest-resolution videos. Then On2 VP6 hit the scene—first as a single codec, then as a codec with multiple options, as you can see in Figure 9-9. Later, as we've discussed, On2 announced—but never shipped—VP8, which was ultimately bought by Google and open-sourced as a component of WebM.

When producing for Flash with VP6, you typically produce an FLV file, which can contain VP6, Spark or even H.264-encoded video, though the F4V extension is increasingly used for H.264-encoded files for Flash. The audio codec for the FLV, however, is almost always MP3, except in the case of some live tools, which use the Nellymoser codec.

VP6-S/VP6-E

The most important VP6-related parameters are VP6-S and VP6-E, presented in Figure 9-9. Basically, VP6-S stands for "Simple," which means slightly lower quality, but an easier-to-decode file. I equate VP6-E to "Excellent," which means the best possible quality, but a harder-to-decode file. If your encoding tool doesn't offer the VP6-S/E option, you're producing VP6-E.

Figure 9-9. The two most relevant VP6 configuration options.

In my comparative HD tests, I found some quality difference between VP6-S and VP6-E, but it was not significant. You can see this in Figure 9-10.

Figure 9-10. The VP6-E image does look slightly better, but you wouldn't notice it
without side-by-side comparisons.

On the other hand, the difference in playback frame rate for a 720p file was significantly different on lower-power computers, as shown in Table 9-1. Specifically, on my daughter's iMac, a
2.0 GHz Core 2 Duo CPU, VP6-E played at about 2 fps, while VP6-S played at full speed.

	VP6-E 2 Mbps	VP6-S 2 Mbps
iMac, 2 GHz Core 2 Duo		
Processor (percentage of two CPUs)	87.5%	91.7%
Drop audio	No	No
Estimated frame rate	2 fps	Full
HP xw4100, 3 Ghz P4 with HTT		
Processor (percentage of overall CPU)	54.6%	51.5%
Drop audio	Yes	No
Estimated frame rate	20 fps	26+ fps

Table 9-1. Playback of VP6-E vs. VP6-S for a 720p file.

On an older Windows-based HP workstation with a 3.0 GHz Pentium 4 CPU with Hyper-
Threaded Technology, VP6-E played at about 20 fps, while VP6-S played at 26 fps or higher.
More significantly, however, audio dropped during VP6-E playback, which is very noticeable.

I didn't see similar discrepancies between VP6-E/S at 640x480 resolution, which leads to the
following simple rule: When you're producing SD video with the VP6 codec, use VP6-E. If

you're producing at greater than 640x480 resolution, use VP6-S. Again, if your encoding tool doesn't give you both options, you're producing in VP6-E, which isn't the best option for HD video.

The only other VP6-specific encoding parameter that you'll consistently see relates to encoding time vs quality, like that shown in Figure 9-11. The impact of this setting varies by tool, but I typically choose the Best setting.

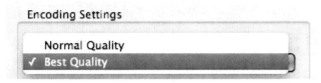

Figure 9-11. Some VP6 encoding tools will give you this quality option; I always go with Best.

How to choose a VP6 encoding tool? Most VP6 encoders produce nearly identical w, so comparative quality isn't an important factor. However, since VP6 is a single-threaded codec, encoding times can be a huge issue. That is, even if you have an 8-core encoding station, unless your encoding tool can render multiple VP6 files simultaneously, it won't encode faster than a single-CPU system. In particular, if you're encoding multiple VP6 files, serial encoders, like Adobe Media Encoder or Telestream Episode, are much slower than encoding tools that can encode in parallel, like Sorenson Squeeze, Telestream Episode Pro (two simultaneous encodes), Rhozet Carbon Coder or Episode Engine.

For example, here are some stats from a recent encoding roundup I wrote for the 2011 *StreamingMedia Sourcebook* (**bit.ly/bestencoder**). On Windows, Adobe Media Encoder, Episode Pro and Sorenson Squeeze all took between 10:51 (min:sec) and 12:36 to encode a single 3:09 file to VP6 format on my 3.2 GHz, 12-core HP Z800.

However, working in parallel, Squeeze rendered eight 3:09 files to VP6 format in 12:27, while Episode Pro took 44:08 and Adobe Media Encoder took 99:07. Mac encoding times for the eight-file test on my 8-core Mac Pro were 20:01 for Squeeze, 49:24 for Episode Pro and 157:12 for Adobe Media Encoder. I retested with Episode Pro 6.1.1 (the original review was version 6.0) and saw no significant difference in quality or encoding speed.

If you have an old copy of On2 Flix Pro, which is now off the market, you can load multiple instances of the program on either the Mac or Windows platforms and encode in parallel as described in **Multiple Instances** in Chapter 10. Otherwise, assuming that you can't part with several thousand dollars for Episode Engine or Rhozet Carbon Coder, Squeeze should be your encoder if you need to produce multiple VP6 files on a tight schedule.

Producing Windows Media Video (WMV)

While Windows Media Video isn't quite as dead as Latin, it's certainly fair to call it a legacy codec, and its usage has dropped dramatically since its heyday back in the late '90s and early 2000s. Still, let's spend a bit of time discussing some encoding-related issues and look at the encoding interfaces presented by the various tools that support WMV output.

The first issue is, What is VC-1 and how does it relate to WMV?

VC-1, WMV and Royalties

Back in July 2007, I asked Microsoft's Ben Waggoner, "What's the relationship between Windows Media Video 9 and VC-1?" Here's his reply:

> This has admittedly been confusing. VC-1 is the Society of Motion Picture and Television Engineers (SMPTE) designation for the standardized codec also known as Windows Media Video 9 (WMV). They're the same thing, technically. You can think of WMV as the name of Microsoft's implementation of VC-1. However, we're increasingly moving toward the use of the VC-1 name to simplify things and avoid confusion.

> That said, VC-1 was designed as a comprehensive codec that scales from HD (shipping in HD DVD and Blu-ray players today) all the way down to mobile devices, including mobile phones. By making VC-1 widely and openly available as a SMPTE standard we were able to achieve one of our primary goals of enabling even greater accessibility and adoption of this advanced codec in the industry. Additionally, a patent pool was created by MPEG LA for VC-1 to provide uniform licensing terms and transparency to third parties.

There are two major takeaways from Ben's comments. First, WMV and VC-1 are the same codec with different names. Second, WMV/VC-1 may be subject to royalties. On the second point, if you create optical discs with VC-1, or use the codec to produce video for pay-per-view or subscription services, you may already owe a royalty. However, if you're using VC-1 to stream free Internet content, here's the relevant provision from the MPEG-LA VC-1 License Summary (**www.mpegla.com/main/programs/VC1/Pages/FAQ.aspx**).

> In the case of Internet Broadcast VC-1 Video (VC-1 Video that is delivered via the Worldwide Internet to an End User for which the End User does not pay remuneration for the right to receive or view i.e., neither Title-by-Title nor Subscription), there will be no royalty for the life of the License.

So no royalty ever for free Internet usage, which is great. On the other hand, since WMV/VC-1 hasn't been top of mind for encoding tool vendors, and because WMV and VC-1 are the same, there's a confusing mix of how the codec is presented in the various encoding tools. For example, the Windows version of Adobe Media Encoder presents the options like this:

Figure 9-12. Adobe Media Encoder presents the various Windows Media codecs like this.

Where Microsoft's Expression Encoder 4, the flagship application for encoding WMV files, presents the codecs like this:

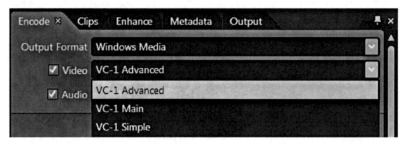

Figure 9-13. Expression Encoder 4 presents the various Windows Media codecs like this.

And Telestream Episode Pro presents them like this:

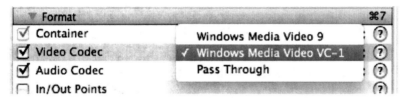

Figure 9-14. Episode Pro presents them like this.

Confused yet? Well, here's the skinny. There are two main codecs within the group: the Windows Media Video 9 codec, which includes the Simple and Main Profiles, and the Windows Media Video 9 Advanced Profile (WMV-AP).

Historically, most web encoding has used the WMV 9 Main profile, with the Simple Profile reserved for CPU-challenged devices like smartphones and MP3 players. The critical differences between the Main and Advanced Profiles relate to support for interlaced formats and other requirements of IPTV and high-definition DVDs.

A couple of key points to note about the WMV-AP codec. First, since the main differences relate to support for interlaced formats, it delivers very little quality benefit for traditional web-based video. That is, you won't see a noticeable increase in quality between a file encoded with the Windows Media Video 9 codec using the Main profile and one encoded with the WMV-AP codec.

Why use the Advanced Profile? Because in the Advanced Profile, Microsoft also exposed a number of encoding options, much like those available for H.264, that are not available with the Main profile. Originally, you had to access these using a registry editing tool called the WMV9 Power Toy, but now these controls are built into several different programs, including Expression Encoder 4 and Sorenson Squeeze (Figure 9-15).

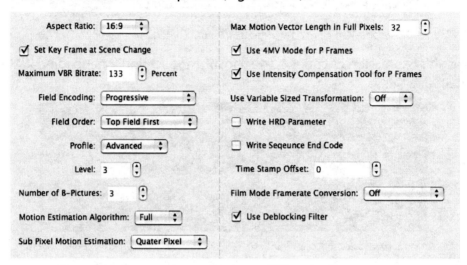

Figure 9-15. Sorenson Squeeze's advanced VC-1 encoding parameters.

I'm all for advanced encoding parameters, but in my experience—at least for files with varying types of content, like my standard test files—these advanced parameters deliver no improvement in quality. For example, in the last round of encoder testing that I performed for the *StreamingMedia 2011 Sourcebook*, Adobe Media Encoder, encoding using the Main profile, produced very similar quality to that of Expression Encoder and Sorenson Squeeze with all parameters set to maximize quality. The only tool that was a little behind quality-wise was Episode Pro.

In addition, while Expression Encoder does a good job explaining what its advanced parameters actually do, Microsoft doesn't tie them to a use case. Beyond some obvious controls, like **B-frames** and closed GOPs (or **IDR Frames**), it's tough to tell when to use which controls or how to configure them. Squeeze provides zero guidance (though I hear some is coming), which is my biggest gripe with the program—relevant documentation is virtually nonexistent.

What about encoding throughput? As you can see in Table 9-2, on Windows, if you're encoding a single file, Episode Pro was the slowest, producing a 3:09 (min:sec) test file on my 3.2 GHz,12-core HP Z800 in 4:33, compared with 3:01 for Squeeze, 2:20 for Expression Encoder and 1:28 for Adobe Media Encoder. Loading multiple versions of Squeeze and Expression Encoder dramatically accelerated multiple-file encoding, with Expression Encoder finishing eight 1-minute files in 4:36, and Squeeze in 7:42. In contrast, Adobe Media Encoder took 15:44, while Episode Pro took 40:12.

On the 2.93 GHz, 8-core Mac Pro, Squeeze finished one file/eight files in 3:45/18:40, while Episode performed the same chores in 7:35/41:36. On both platforms, considering quality and throughput, Squeeze is the top choice. Again, I retested with Episode Pro 6.1.1 (the original review was version 6.0) and saw no significant difference in quality or encoding speed.

	Adobe Media Encoder	Microsoft Expression Encoder 4	Sorenson Squeeze	Telestream Episode Pro
Windows encoders				
One 3:09 file	1:28	2:20	3:01	4:33
Eight 3:09 files	15:44	4:36	7:42	40:12
Mac encoders				
One 3:09 file	NA	NA	3:45	7:34
Eight 3:09 files	NA	NA	18:40	41:36

Table 9-2. Mac and Windows WMV encoding times.

So here's how I recommend approaching Windows Media encoding: First, don't expect advanced encoding controls to translate to better quality. If you have Adobe Media Encoder on Windows (the Mac version doesn't produce WMV files) and throughput doesn't matter, it should be sufficient, even though it offers no WMV configuration options. Second, when you're using a tool like Squeeze or Expression Encoder, rely heavily on the presets—don't spend a lot of time messing with the controls because you'll be flying blind and it likely won't make a difference.

Now onto our last codec: Ogg Theora.

Ogg Theora

Before WebM, Ogg Theora was the only open-source, royalty-free codec available for the HTML video tag. However, its quality was noticeably inferior to that of H.264 as you can read at **bit.ly/ogground1**. Be sure to read all the comments; there are some amusing ones.

However, the launch of WebM orphaned Ogg Theora because all the HTML5-compatible browsers immediately supported WebM, which provides much higher quality than Ogg Theora. If your goal is to support every version of every HTML5-compatible browser, you'd still have to produce Ogg Theora files, but I'm guessing few people who feel that stridently about this issue would actually buy this book. So I'm going to declare victory for this chapter, and move on to the next.

If you are interested in Ogg, the foregoing reference details how I produced multiple versions of test files, while "Ogg vs. H.264—a real-world view" (**bit.ly/oggrealworldview**), encapsulates my feelings about Ogg in general.

Conclusion

Now you know how to produce files in multiple formats; in the next chapter, you'll learn how to produce them more quickly, using one of two techniques to accelerate encoding on multiple-core systems.

Chapter 10: Distributing Your Video

By this point, you know all you need to know about encoding your video. Now it's time to distribute it to the world. You have three basic approaches.

First, you can encode the video files yourself, create the necessary player and all the links, and upload the files to your own website or a web server leased from a third-party provider like Amazon. As long as viewing numbers stay fairly modest, this approach should work from a technology standpoint, though it may not be the best way to get maximum eyeballs for your videos.

The second alternative is to host your videos on a free, user-generated content (UGC) website like Vimeo or YouTube. These UGC sites relieve you of the encoding and player-creation chores and assume the task of hosting and distributing the video for you. You can still embed the video on your own website, but by offering your video on a UGC site, you also expand the number of potential viewers, which can help from a marketing perspective. However, there are some negatives to consider, as well as some benefits that are only possible via the third alternative.

That alternative is to use a fee-based service to host and distribute your videos for you. Multiple online video platform (OVP) vendors offer hosting, encoding, customizable players and detailed statistics to help you maximize the effectiveness of your video. They help distribute your video to other sites to acquire more viewers. They also can provide interactivity that lets viewers click the video to advance to the next step in the sales cycle, as well as other features. Even a few short months ago, these types of features would have cost hundreds of dollars per month. Today, however, depending upon the amount of video you distribute, you can sign on for less than $20 a month, with one service offering unlimited video views for less than $50 a month.

This chapter will review the costs and the benefits of all three alternatives. While there is no one-size-fits-all solution, any business seeking to truly leverage the value of its video should at least be familiar with the benefits of the second and third alternatives. Many businesses will find that an amalgam of these two is the best option of all.

Before jumping into the three options, let's look at the roles related to video distribution.

Understanding the Roles

Table 10-1 delineates the functions related to successful video distribution and how they're divided based upon the three distribution options. I've added a fourth—using a cloud-based streaming server—because it's gaining in popularity.

	Encode	Server	Feature Support	Platform Support	Player Creation	Syndicate (get viewers)	Deliver
Do it yourself	You	You	You	You	You	You	You
User generated content (UGC)	UGC	UGC	UGC	UGC	UGC	UGC	UGC
Online video platform (OVP)	OVP	OVP	OVP	OVP	OVP	You	OVP
Cloud server	You	Cloud	You	You	You	You	Cloud

Table 10-1. The roles related to video distribution.

The encoding role is what you've learned in this book; whether you want to do it on a daily basis is up to you. The server role relates to installing and maintaining the streaming server. While a streaming server isn't essential for simple streaming, it is for advanced features like digital rights management, comprehensive analytics and some adaptive streaming formats.

Feature support relates to these ancillary features. As we'll discuss, when choosing an OVP, you'll compare dozens of features, like content management systems, various syndication and monetization models, and support for adaptive streaming. If you use a third-party service, it will develop and implement these features; if you DIY, you DIY. Ditto for platform support. If you're rolling your own distribution and want to support iOS and Android devices, you have to build it yourself. Use a UGC or OVP site, and it does the heavy lifting. Sound far-fetched? Maybe, but by using YouTube to host its videos, IBM was one of the first corporate sites to post videos that played on an iPad.

Speaking of that, player creation is a huge feature supplied by OVPs and UGC sites. Creating a simple player is a breeze, but a simple player may not be sufficient. As I discuss in "Is Your Video Player as Good as Your Content? (bit.ly/gezmZg), to maximize the impact of the video, a fully featured player needs to:

- Foster viewer engagement via features like number of viewers, ratings and comments

- Link to social networking sites

- Enable sharing options like linking, embedding and emailing

- Contain links to other content that viewers might want to watch

- Allow viewers to easily find the content that they want to view.

For example, Figure 10-1 shows Kohler's video player, which is provided by OVP Brightcove.

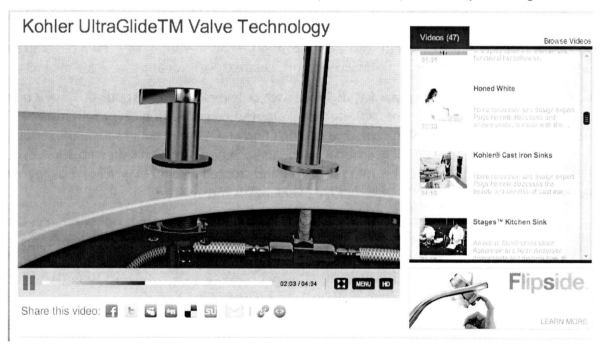

Figure 10-1. The Kohler video player, courtesy OVP Brightcove.

Beneath the player are links to Facebook, Twitter and other social media sites, plus links for embedding and emailing. On the right is a playlist containing related videos that might prove enticing to the viewer, which makes the site much more likely to retain the viewer's interest, or "sticky." Most UGC sites and OVPs enable most of these features, but to develop them yourself might take days or even weeks of programming time.

Finishing up with Table 10-1, syndicate means getting additional eyeballs for your video. Some UGC sites, like YouTube, do this inherently, as you doubtless already know. While OVPs don't typically have lots of viewers surfing around looking for videos to watch, as we'll discuss, some OVPs can automatically send your videos to sites like YouTube or Metacafe to garner additional attention.

The final role is delivery. If you host the videos on your own website, it's OK for a few dozen users, but your server is likely not optimized for high-volume video delivery. In contrast, most UGC and OVP sites have contracts with content delivery networks (CDNs) to ensure high-quality delivery of high-bit-rate streams.

Distributing Your Own Videos

As we've seen, if you distribute your own videos, you perform all roles from encoding to delivery. Certainly you can spread the load by licensing a server in the cloud, which saves you the server installation step and should take care of delivery as well. Or you could hire a CDN to distribute your videos. But you would still have to configure and maintain the server, develop all features, develop support for additional platforms, create the player and chase the eyeballs.

In my experience, for small companies, this approach only makes sense when you have a need that can't be satisfied by a UGC or OVP site. For example, one consulting client sells music-training videos on a subscription basis and couldn't find an OVP that could satisfy the requirements of its revenue model. So it licensed a Flash Media Server from Amazon Web Services, programmed the required features into the server and created its own player. For most general-purpose users, however, a UGC site or OVP likely makes more sense.

If you distribute your own videos, the most critical decision is the streaming server that you buy or license. Remember, you don't absolutely need a streaming server for simple progressive delivery, but you will for advanced features like advertising support, syndication and server-based content management. If you do buy or license a server, obviously you should choose one that supports all of your current and short-term future platforms.

For most producers, this list should start with Flash, and any of the Adobe Flash Media Server products can fit the bill (adobe.ly/flashserver). However, if you also want to target Silverlight, iOS or Android viewers, consider the Wowza Media Server 2 (www.wowzamedia.com) or RealNetworks Helix Server (bit.ly/realhelix).

If you do choose a cloud provider of the chosen server (both Flash and Wowza servers are available in the cloud), check to see which CDN it uses to deliver its videos. You'll want a reputable CDN like or Level 3 to ensure the high-quality delivery of your videos.

Adding Video to Your Website

If you're adding video to your website without a streaming server, the process is simple, though it does require some technical know-how. For example, step 1 is uploading the file to your website, which is typically accomplished with an FTP program like freeware FileZilla, shown in Figure 10-2.

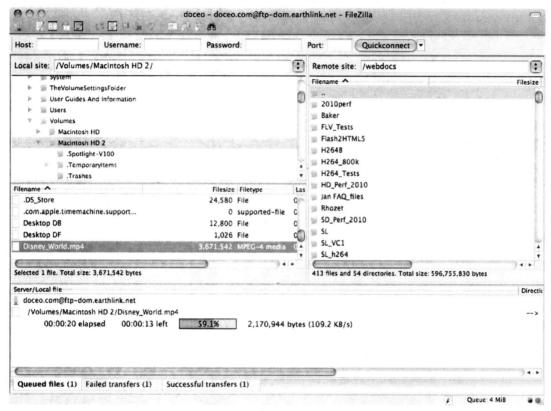

Figure 10-2. Uploading Disney_World.mp4 to my www.doceo.com website.

Once the video file is uploaded, you can email a link to a viewer, who can play the video file by either clicking the link in the email, or copying and pasting the link into his or her browser. For example, if you go to www.doceo.com/Disney_World.mp4, you'll play the video file that I uploaded in Figure 10-2.

If you're going to email a link, things will go easier if you don't leave any spaces in the file that you upload, since that tends to drive some browsers crazy—that's why there's an underscore between Disney and World in the file. Also to avoid sending the wrong URL for the link, I first enter the link in my browser, make sure it pulls up the right file, and then copy it into my email. Note that things like capital letters matter, so it's easy to type in the wrong link.

Linking and Embedding

If you want the video to play within a web page, you have to embed the video into the web page. The technique and language for doing this varies according to the technology that you're using.

With HTML5, for example, the syntax is simple. Here's the embed code for a VP8 file that I used on my own website at www.doceo.com/HD_WEBM.html:

```
<video src="http://www.doceo.com/HD_800.webm"
       type='video/webm; codecs="vorbis,vp8"'
           controls
           autoplay>
```

Figure 10-3. HTML5 embed codes are very simple.

You can find a good explanation of the controls available for HTML5 video at www.w3schools.com/html5/tag_video.asp. However, note that simply adding the tag to an HTML page on your site may not be sufficient—you may also have to tell your web server how to deal with this new type of video file, which is typically called a MIME type. For example, to get any file to play via HTML5 on my website, I had to add the following MIME types to the .htaccess file on the www.doceo.com website.

```
AddType video/ogg  .ogv
AddType video/ogg  .ogg
AddType audio/ogg  .oga
AddType video/mp4  .mp4
AddType video/webm .webm
```

Figure 10-4. MIME types on the www.doceo.com website.

If you need help with these MIME issues, particularly with an Apache web server, click over to diveintohtml5.org/video.html and search for "MIME type" on that page.

Since only about 50% of the installed base of browsers are HTML5-compatible, you'll probably want your embed codes to direct older browsers to fall back to Flash or other plug-in based playback technology. This is a bit far afield for me, but Google "HTML5 video tag" and "Flash fallback" and you'll find some great resources.

Embedding Flash Video

Embedding Flash videos is complicated because there are multiple options for displaying Flash video, and all use complex embed codes. I don't try to hand-code Flash embed codes any longer; I usually create the web pages in a tool like Adobe Dreamweaver or Adobe Flash Catalyst. You can see a tutorial on using Adobe Flash Catalyst at bit.ly/ozercatalyst.

The only frustration is that these tools typically won't input MP4 or even F4V files (in the case of Dreamweaver), even if the video is encoded in an H.264 format that the Flash Player can play. The simple fix is to manually change the extension to .flv in Windows Explorer or Finder so the program will recognize the file. Flash Player will still be able to play the file, so changing the extension shouldn't cause any problems.

Now let's look at using a UGC site to distribute your videos.

Distributing via UGC Sites

When you work with a UGC site like YouTube, first you set up your free account and then you upload your video to the site. UGC sites will display the videos on their own pages and provide an embed code that you can use to display the video on your own web pages.

Uploading is straightforward, so I'll trust you to work through that one yourself. Embedding is also simple; click Embed to open the embed codes, choose your options, then click inside the embed code box and copy the embed code to your clipboard.

Figure 10-5. Grabbing the embed codes from YouTube.

Typically, the most important option is size, because you'll want the video to fit within the horizontal boundaries of your web page. My blog is 700 pixels wide, which is why I used the 640x360 option for this video.

The only tricky part is pasting the embed code into your web page. Specifically, remember that you're pasting HTML embed codes into your site, which you must do directly in the HTML editor provided by your authoring program or content management system. For example, I use the Interspire content management system for *StreamingLearningCenter*, and it has an HTML editor that I access via the circled icon on the upper right of Figure 10-6. I don't paste the embed codes into the normal text entry area; I paste them into the HTML Source Editor you see in Figure 10-6.

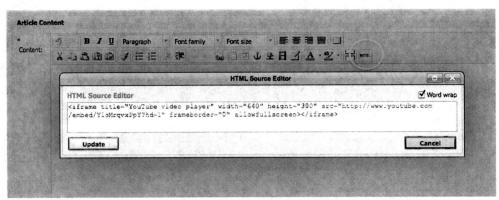

Figure 10-6. Accessing the HTML editor in my Interspire content management system.

It's a bit easier to see in the WordPress blog shown in Figure 10-7. Basically, if you ever see the HTML codes themselves in your website text, you should know that you've copied them into the wrong editor.

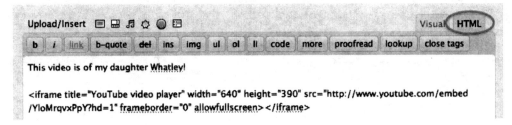

Figure 10-7. Accessing the HTML editor in WordPress.

Note that this embed procedure is the same whether you're pulling embed codes from a UGC site or an OVP. Either way, you get the embed code and paste it directly into the HTML editor on your website. If you'd like to see a tutorial on uploading to and embedding pages from a UGC site, click over to bit.ly/uploadembed.

Choosing a UGC Site

Now that you know how to use a UGC site, let's focus on which one to choose. There are two aspects: which sites do you want to use to reach potential viewers, and which do you want to use to host the videos that you embed in your own site?

Chasing Eyeballs

If it's attention that you want, it's hard to go wrong with YouTube. If you scan YouTube's channels, you'll note some pretty interesting names with some impressive view counts. Etrade has about 27,000 subscribers with more than 52 million video views; the White House has more than 125,000 subscribers with more than 43 million video views; IBM has 4,058 subscribers and about 2.3 million video views; Intel has 17,963 subscribers and more than 17 million views; Nike Football (which is really soccer) has 86,673 subscribers and more than 75 million video views, which is topped by Taylor Swift's 549,133 subscribers and 89 million video views.

Clearly, exposure on YouTube isn't for everyone. But for many products and services, it's an inexpensive route to a vast audience you probably couldn't reach by posting videos on your own website.

The other thing about YouTube is that it isn't particular. That is, so long as you're not violating someone's copyright or uploading pornography or other "bad stuff," you're free to sell and market your products or services as you wish. This isn't the case with many other sites, particularly Vimeo, which says:

> You may not upload commercials, infomercials, or demos that actively sell or promote a product or service. Businesses may not use Vimeo to externalize their hosting costs. Vimeo (including Vimeo Plus) is not a business service.

When chasing eyeballs, my theory is that more is always better. In this regard, you should be aware of a website called TubeMogul that offers a free service that distributes your videos to multiple sites from a single upload. Target sites include UGC sites like YouTube, DailyMotion, Metacafe and Viddler, social media sites like Facebook and MySpace, and OVP sites like Brightcove and Viddler.

Operation is simple; you upload your video, add metadata, choose the target sites that you want TubeMogul to deliver to and enter the required login information (Figure 10-8). TubeMogul does the rest. I haven't used TubeMogul in the past, but now that I'm producing more tutorial videos, I'll certainly be using it to chase eyeballs going forward.

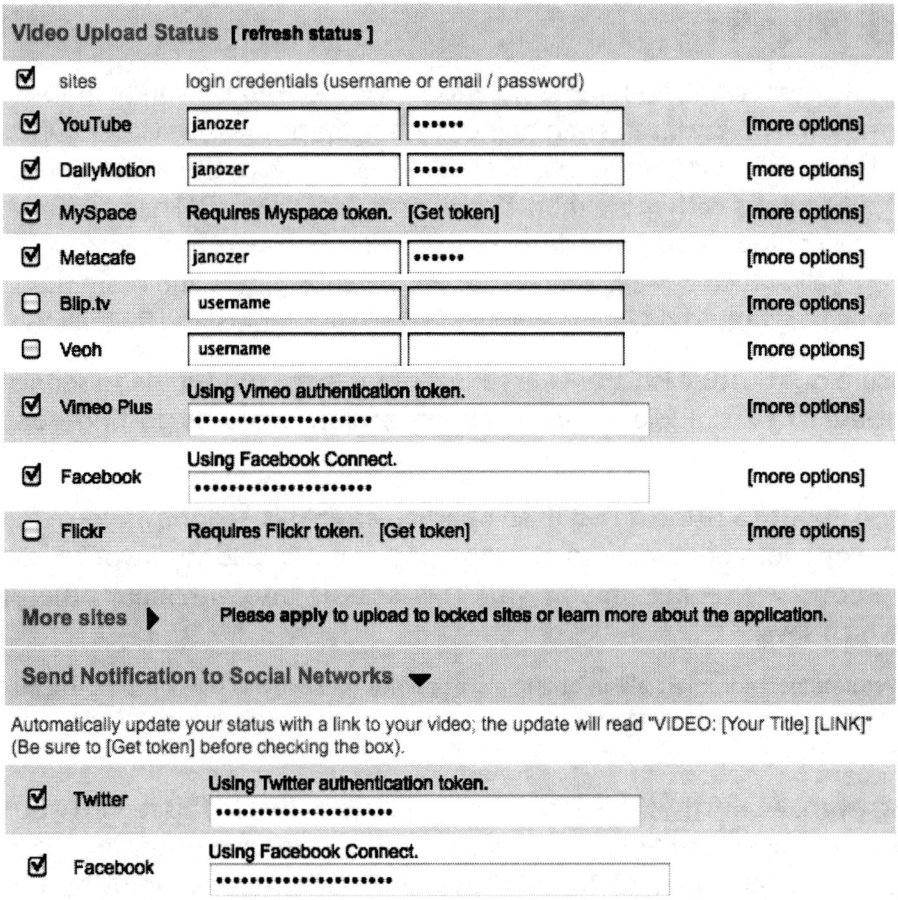

Figure 10-8. TubeMogul syndicates your videos to multiple sites for free

Choosing a UGC Host for Your Videos

Eyeballs are one thing, but which UGC site should you use to host the videos that you embed in your website? That's a different kettle of fish.

As mentioned above, whatever site you choose, make sure you're not violating its terms of service. I use Vimeo to serve videos on my website, but most of my videos are encoding-related tutorials or other demonstration videos. If I started adding advertisements to these videos, that would clearly violate Vimeo's terms of service, so I'd choose another service.

In early 2011, I reviewed five UGC sites plus WordPress' VideoPress service to determine their suitability for hosting videos embedded in another site (see "Choosing a UGC Site," bit.ly/chooseugc). Table 10-2 summarizes the characteristics that I considered. To be fair, VideoPress does cost $59.97 a year, and I compared it with the free services offered by the

UGC sites. Still, if you have WordPress blog, it's not a lot of money and it's a service you should consider.

Feature	YouTube	Vimeo	Viddler	DailyMotion	VideoPress	Metacafe
HD	Yes	One/week	No	Yes	Yes	Yes
Quality HD	Excellent	Very good	NA	Good	Excellent	Good
Quality SD	Excellent	Excellent	OK	Fair	Excellent	Good
Customize size	Yes	Yes	Yes	Yes	Two sizes	Three sizes
iPad/iPhone	Yes	Yes	Didn't work	Yes	Not really	No
USP	Eyeballs	Quality	Interactivity	Advertising	WordPress	Eyeballs

Table 10-2. Comparing UGC sites for serving as embedding host.

Starting at the top, not all video has to be HD, but it's a nice bonus when your sources are HD. If you need HD video, Viddler is out unless you want to pay a few hundred dollars a month. For top-quality HD, or SD for that matter, go with YouTube, Vimeo or VideoPress.

On the player side, being able to customize size is critical to optimal embedding, and I favor sites that let you choose the exact horizontal resolution, not just between two or three sizes. The value of iOS compatibility speaks for itself.

Finally each site has its own unique selling proposition (USP), positive or negative. With YouTube, it was eyeballs on the YouTube site, while Vimeo offers great quality and access to its artsy community and Viddler offers interactivity, or the ability to create links in the videos that viewers can click to access desired sections. DailyMotion's USP—advertising—is a negative, since when you embed videos from DailyMotion on own your site, third-party advertisements play before your own videos do. VideoPress has great quality and runs from within your WordPress dashboard, while Metacafe delivers a diverse group of potential viewers.

Overall, while other sites may be valuable for the eyeballs that deliver, YouTube delivers an unbeatable combination of high-quality video, player customizability, millions of potential viewers and commercial-friendly terms of service. Unless you've got an aesthetic aversion to YouTube, you should strongly consider using YouTube to host the videos that you embed in your own site—though if I had a WordPress blog, I'd use VideoPress because it's so convenient.

That's choosing a UGC site; now let's move on to choosing an Online Video Platform.

Online Video Platforms

Let's take a short look back at the history of streaming delivery to help explain the value proposition of the current crop of Online Video Platforms (OVPs). Back in the late 1990s, when streaming became mainstream, the biggest category of service providers was content delivery networks (CDNs). These companies would store your videos on their servers, distribute them on their networks and ensure their smooth distribution to the ultimate viewer far and wide.

The CDNs and their services spawned a range of third-party service providers offering ancillary services such as player interactivity, viewer statistics, syndication of your video to other websites and hooks to advertising-related services—the latter two so you could make money from your video content. Over time, these services were merged into a single, consolidated product offering by companies such as Brightcove and Ooyala.

As you would suspect, these companies targeted media companies with high stream counts and the distribution budgets to match. There are only so many media companies around, though, so the third-party service providers' focus inevitably turned to corporate users, from small to large.

As their focus changed, so did minimum pricing, which transitioned from low four-digit figures to pay-as-you-go. Sign-up procedures, which initially involved much paperwork and significant up-front financial commitments, have also become much more streamlined. One vendor, Ooyala, claims that you can download its Backlot software and start streaming in less than 15 minutes without speaking to an Ooyala employee or making any long-term financial commitment. Pricing varies dramatically from vendor to vendor, so do your homework in that department early in the selection process.

However, though getting started with an OVP is simple, choosing one isn't. There are dozens of companies entering and exiting this market—some via consolidation, some via bankruptcy.

As you read about the introductory features discussed next, recognize that unless otherwise indicated, many are generic and not unique to the company identified. The purpose of this section is to identify the high-level features and benefits of using an OVP, not to tie features to specific providers.

Basic OVP Capabilities

You work with an OVP the same way you work with a UGC site: you upload the file to the service; it encodes and supplies a player and embed code. Regarding the player, most vendors offer the ability to create a branded player with all the normal playback controls and embedding and email options, if desired. Another common feature is the ability to embed a single player in a page and create a video library for viewers to click through, such as the Kohler library shown in Figure 10-1.

Be sure to check the extent to which you can customize the player and video libraries presented within the page, and note how easy or how difficult these features are to customize. For example, some OVPs offer drag-and-drop widgets that let you easily create players and video content windows in a variety of shapes and sizes, along with a comprehensive programming interface for customization. Brightcove offers four levels of player and video window customization, from simple enabling and disabling of player-related features to a programming interface for custom programming or dropping in third-party plug-ins.

Also determine if you can create different players for various pages on your website and custom players for customers or others to distribute your videos. For example, since Kohler sells solely through distributors, it has produced extensive libraries of streaming content to reach potential customers directly. Kohler makes these libraries available to its distributors with a custom player that reflects both Kohler's and the distributor's branding.

- **Resolution and data rate flexibility.** You're hiring an OVP to deliver video to your website at the resolutions and data rates desired by your web designers, not those selected by the OVP. Make sure that your OVP supports all the resolutions and aspect ratios that you plan to use on your site, at the data rates necessary to make them look good.

- **Content management.** All OVPs supply basic content management capabilities that typically include the abilities to tag media for search and retrieval, to select preview images or thumbnails for the videos, to incorporate multiple videos into a playlist for sequential playback and to incorporate the uploaded videos into your own content management system.

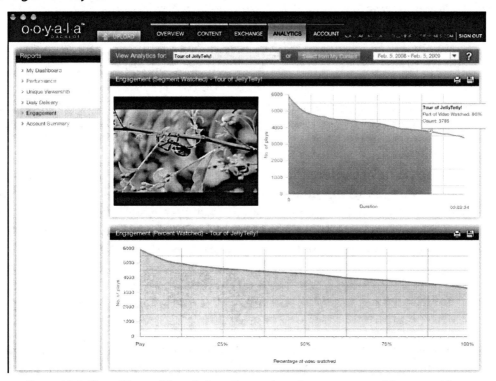

Figure 10-9. These "drop-off" analytics tell you when viewers stop watching your videos.

- **Analytics.** Once you make your videos available, you'll want to know how and where your videos are played. At the very least, most OVPs detail video views and report download bandwidth and details like the country and domain of your viewers. Some offer the ability to download CSV files so you can further analyze this data in Excel. If you're a current user of analytics programs from providers such as Google, Omniture, or Visible Measures, look for the ability to integrate your data into these packages.

Beyond these basics, many OVPs also present true viewer analytics that allow you to identify patterns within the statistical data. One common and exceptionally useful feature are "drop-off" statistics that identify where viewers stop watching the video (Figure 10-9). For example, if you have massive drop-offs in the first 30 seconds or so, you have the data to back up recommendations such as ditching the CEO's greeting or the 20-second introductory animated collage. If few viewers make it through to the end of your 4-minute product demo, you know that you need to get to the point more quickly or, perhaps, present that video later in the sales cycle. If you're displaying advertisements in your videos, you'll be able to tell where they have the most impact, and where they simply drive viewers away.

Beyond these basics, there are many advanced features that you should consider when choosing an OVP.

- **Supported platforms.** Virtually all OVPs have Flash-compatible players, but make sure they're streaming with H.264, not VP6. If you want Silverlight, you'll have to dig harder since relatively few OVPs support Silverlight (and again, make sure the service streams H.264, not WMV). In the short term, the most important aspect of HTML5 support involves reaching iOS devices, which you can do two ways: via the browser or via an iPhone or iPad app. If an app is in your plans, find out if the OVP has a software development kit (SDK) that can assist your efforts. Also ask if iOS support is single-stream or adaptive.

  In addition to iOS devices, determine the OVP's plans for Android, BlackBerry, HP webOS and other relevant platforms, again asking whether support will be browser-based, app-based or both, and if adaptive streaming will be supported. Ask the OVP about its scheme for browser detection, which is necessary to direct mobile visitors to the proper stream. Does the company have logic you can use to accomplish this, or do you have to develop your own?

  If you're a sandal-wearing, tree-hugging HTML5-fanatic, ask about WebM support. Not that I'm being judgmental, of course.

- **On-demand, live, or both.** If live events are part of your content mix, determine if the OVP handles both on-demand and live events. If live is a major component of your strategy, check out **Chapter 11, Streaming Live Events**. If the OVP does support live, determine the ancillary formats supported (PowerPoint? Screens from other applications?), and social media components (chat? polls?), and ask if you can get production support if you don't have your own camcorder, encoding gear or in-house production talent.

- **Single stream or adaptive.** By 2011, if an OVP doesn't offer adaptive streaming, it is far behind the times. Ask whether the OVP can transmux the streams you input to distribute them to multiple formats, or whether you need to send it a discrete stream for each format. Obviously, the former approach is desired.

- **Supported protocols.** As you learned in Chapter 7, Protocol (RTMP vs HTTP), there are multiple protocols for distributing your on-demand and live video. Though RTMP is more proven at this point, many experts feel that HTTP enables higher-quality, higher-bit-rate streams. If you're streaming in Flash, ask which protocol the OVP supports, which hopefully will be both. Also ask if the OVP supports multicasting and peer-to-peer delivery, which were introduced by Adobe with Flash Media Server 4. Most other formats, including Silverlight and Apple's HTTP Live Streaming for iOS devices, are HTTP.

- **OVP Integration.** You can use your OVP like a third-party service, uploading and managing videos on its site, or you can integrate its functionality into your content management system, which simplifies adding videos to your website. For example, Figure 10-10 shows how VideoPress integrates into a WordPress site, which is via the circled icon in the Figure. If you have multiple users contributing to your website, this level of integration requires much less training and support than a third-party service.

If third parties will be uploading videos to your site, you also want the pages they'll see to reflect your own branding. If these issues are important to you, ask the OVP about alternatives for integrating the OVP functionality into your content management system and website. For example, many OVPs offer plug-in support for common content management systems, like WordPress, Drupal, Joomla and others.

Figure 10-10. VideoPress integrates extremely well into WordPress blogs.

- **Support for your business models and stream security.** At some level, all producers need to monetize their videos, whether via third party or in-house advertising, or some other technique. So be sure that the OVP supports your intended models. If you'll be serving ads, make sure the OVP supports the IAB Digital Video-related advertising formats you plan to use (for more on IAB advertising formats, see bit.ly/advertformat). Also ask if the

OVP will support the advertising networks that you plan to work with. If you're not serving third-party ads, ask about intro and outro videos that you can use to market your own products and services.

If you'll be selling your content, be sure that the OVP supports your intended model, whether subscription, pay-per-view or other hybrid model. The flip side of these payment models is securing your stream, so non-paying viewers can't access them. Security techniques can include access control through authentication, limiting access by domain restriction and geo-filtering, stream encryption, and SWF verification. If you need to secure your streams, these aspects will be a major focus for you.

These are the issues that should be addressed by all producers seeking an OVP; following are some advanced issues that you may also want to consider.

Other Considerations

- *Who owns the code?* Some producers prefer open-source alternatives, like that offered by Kaltura, because it allows them to customize their platform and provides protection if the OVP goes out of business.

- *Syndication.* We talked in the UGC section about syndicating your videos to various UGC sites. If this strategy is important to you, determine how easily the OVP can syndicate your videos to YouTube and other relevant UGC targets.

- *User-generated content.* If your monetization scheme depends upon user uploads, check how easily your OVP enables uploading and converting user-generated content.

- *Interactivity.* Interactivity lets viewers click the video window to make something happen, for example, playing another video, jumping to a different webpage or retrieving a PDF file. If you're integrating video into a sales cycle, it's a great way to present the call to action whether it's, "Click here to download a contract," or, "Click here to contact a sales representative."

For example, in Figure 10-11, viewers of the video can click the hot spot shown to purchase the sniper scope reviewed in the video. I spoke with the owner of Sniper's Hide, who reports clickthroughs to the seller's website in the 4-to-6% range with up to 30% clickthroughs for informational (as opposed to sales-oriented) links. For example, in his training videos, he may make pants, boots or other equipment clickable so viewers can view the make and model. If you're really trying to foster viewer response, interactivity is a feature worth considering.

SWFA Super Sniper Scopes 3-9X42 FFP & 10X42 HD

Tue, 10/20/2009 - 18:00 — Lowlight

This is a Sniper's Hide Interactive Video Review of the SWFA Super Sniper Scopes. The Front Focal Plane, Mil Adjusted Mil Dot Reticle 3-9X42, and the Mil Adjusted, Mil Dot Reticle 10X42 HD

Figure 10-11. Interactivity in the Snipers Hide website.

- **Review-and-approval workflow.** Sorenson 360—the OVP from Sorenson Media, the developer of the market-leading Squeeze encoder—has implemented a review-and-approval workflow that lets video producers upload a video and send a link to the client to review the video and make comments. If you frequently produce video for third-party approval, this can really speed the process.

- **Client-side encoding/upload acceleration.** The first is a feature of Sorenson 360, where you can use Squeeze to encode the file before uploading. If files are very large, or upload connection speeds very slow, this can dramatically increase how quickly you can get files posted (and reduce upload time-out and other errors). If client-side encoding isn't an option, ask if the OVP uses upload acceleration technology like that enabled by Aspera technology, which is said to speed uploading by as much as 20 times. Brightcove is an example of an OVP that offers Aspera-based upload acceleration.

- **Where do you want your OVP to live?** Most users will want to access their OVP as a third-party service, though if your organization is large enough, you may want to license the software to bring it in-house, or run it in the cloud. Not all OVPs offer this flexibility, though Kaltura does.

There are literally hundreds of features touted by the various OVPs, but hopefully I've covered the majors. In terms of the decision-making process, following is the procedure used by Bruce Colwin, a serial entrepreneur who has selected multiple OVPs for his various projects.

Choosing an OVP: A Six-Step Process

Bruce Colwin currently works with *LegalMinds.tv*, and this is the process he used to choose an OVP for this site. His first step was to visit **www.vidcompare.com**, a comparison service for decision makers seeking to chose an OVP. You can read a *StreamingMedia* review of VidCompare at **bit.ly/vidcompare**.

Briefly, VidCompare lets you search for OVPs using categories such as Delivery and Payment Structure and provides basic information such as a feature list, a pricing structure, and the date the company was founded. Even though he was familiar with the OVP market, Colwin found VidCompare useful because the comprehensive listing identified candidates that he shouldn't miss. He estimated that he spent about 6 hours total on the VidCompare site before making his decision.

After identifying potential options on VidCompare (which was step 1), Colwin used information on VidCompare and the vendor sites to weed out services that didn't fit his business or pricing model (step 2). Then, he read the company information and any user reviews on VidCompare and polled his business connections to see if any of them had any relevant experience with the remaining candidates (step 3). Next, he contacted the candidates to ask any remaining questions and to make sure that he understood their pricing structures (step 4).

Then, he requested trial accounts with the final candidates (step 5). Over a period of a few weeks, he uploaded content, tagged it, embedded the video into browsers and played it back on multiple computers. During this process, he found that one potential vendor had problems with WordPress that could have been a huge issue with some of his potential clients. With another final candidate, he found the pre-sales service lacking, which soured him on the prospect of its after-sales service. After the trials, Colwin consolidated his findings regarding quality, usability, pricing and customer service, and then made his decision (step 6).

Certainly, I would take all the steps that Bruce recommends, plus ask to talk to reference accounts that are currently using the service. While the OVP will obviously point you to happy customers, I doubt the references would lie to you when you ask about the service. In particular, I would ask how quickly the OVP fixes bugs and other problems, and its to new feature requests.

OK, that's all on choosing an OVP. Now let's focus our attention on encoding for uploading to a UGC or OVP site.

Encoding for Upload

We spent most of the early part of the book learning how to encode video to final form for distribution to your viewers. All UGC sites re-encode your video, so when you're uploading to a UGC site, you're uploading for re-encoding. Some OVPs will distribute the files that you upload, but many re-encode, so you should ask about this before you start uploading files.

Recognize that when you're encoding for re-encoding, you're trying to provide the highest-quality starting point that can be reasonably uploaded to the UGC site/OVP. If you're getting tired of reading, I've created a video that outlines the factors you should consider when encoding for upload, which you can access at bit.ly/encode4upload. As you would guess, the content of this chapter will follow that video pretty closely.

Basic Considerations

Most UGC sites have a maximum upload limit, which can range from 100 MB to 2 GB. So you'll want to know that number first, and make sure that your file doesn't exceed that limitation.

With SD files, I always scale and deinterlace before uploading. That is, if I shoot in SD DV format—say 4:3 aspect ratio—the resolution is 720x480 and the file can be interlaced or progressive. Before uploading, I would scale to 640x480 and deinterlace (if necessary) during the encoding process. For a 16:9 DV file, I would scale to either 640x360 or 848x480 and deinterlace during output.

	SD: 4x3	SD: 16:9	720p	1080p
Resolution	640x480	640x360/ 848x480	1280x720	1920x1080
Scan	Progressive	Progressive	Progressive	Progressive
Data rate (target/max)	5/10 Mbps	5/10 Mbps	10/15 Mbps	20/30 Mbps
Codec	H.264	H.264	H.264	H.264
Profile	High	High	High	High
Entropy coding	CABAC	CABAC	CABAC	CABAC
Bitrate control	2-pass VBR	2-pass VBR	2-pass VBR	2-pass VBR
Key frame interval setting	10 seconds	10 seconds	10 seconds	10 seconds
B-frame/Reference frame	'3/5	'3/5	'3/5	'3/5
Audio channels	Native	Native	Native	Native
Audio codec	AAC	AAC	AAC	AAC
Audio data rate	64 mono/ 128 stereo	64 mono/ 128 stereo	64 mono/ 128 stereo	64 mono/ 128 stereo

Table 10-3. Encoding for upload recommendations.

I always encode files for upload with the H.264 codec at high data rates, following the guidelines shown in Table 10-3. Typically, I encode using the same encoding parameters that I use for maximum quality (High profile, CABAC, three B-frames, five reference frames) though at the high data rate recommended, these parameters probably have minimum impact on quality. In addition, I always use two-pass VBR encoding, with a key frame interval of 10 seconds.

For audio, I recommend uploading the number of native channels in the stream: one channel if mono, two if stereo. I encode to AAC format at 64 kbps for mono, 128 kbps for stereo. Most of my work is simple speech, though; if you're recording the London Symphony Orchestra (or some reasonable facsimile thereof) you might try encoding and uploading the audio component at higher rates.

Table 10-3 consolidates these recommendations into a more usable form. To be honest, though I've used these parameters extensively, I haven't researched them that thoroughly—my basic assumption being that at these data rates, minor adjustments to the H.264 encoding parameters make very little difference. If you're about to start a long relationship with an OVP, I would ask for its recommendations. If it has none, I would try a couple of short files using the recommendations in Table 10-3, and tweak them if necessary.

Conclusion

Now you know how to distribute your on-demand videos via UGC and OVP sites. Next up is choosing your gear and services for live events.

Chapter 11: Streaming Live Events

Though many of the concepts are the same, live streaming is a different animal than on-demand streaming, with different products and service providers. In this chapter, you'll learn the basics of producing a live event, and then how to choose the best encoder and service options for encoding and distributing your live event streams.

Introduction

At a high level, streaming live events comes down to four components, as shown in Figure 11-1: camera, encoder, streaming server and distribution network. You set up the camera, choose and configure the encoder, direct the streams to the server and send that feed to your distribution network, and you're streaming away.

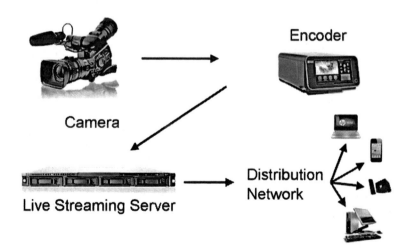

Figure 11-1. The four components of live event streaming.

To link the video to your website, you'll get an embed code from your streaming server or service provider that allows you to link the live feed into your web page. Or if you're hosting the server yourself, you'll create the web page with the necessary links to the live stream. Just so you know, I'm going to walk through the production of a live event at the end of this chapter that will pull these pieces together in a much more visceral way.

I'm not going to discuss choosing a camera in this chapter, though your video source can be anything from a simple webcam to a sophisticated three-CCD professional model. My sole advice here is that better equipment produces superior results and you should always shoot progressive. As for the other components, that's where we're going to dig in, starting with the encoder.

Choosing an Encoder

For most live events, the stream that you capture onsite and send to your streaming server is the stream that's actually distributed to your viewers, though there are some exceptions. For example, Sorenson Media recently launched Squeeze Live, a cloud-based service that transcodes a single stream from a live event into multiple streams for adaptive delivery. This limits the number of encoders you need onsite and lets you distribute adaptive streams with limited outbound bandwidth.

In addition, some very high-end broadcasters send an HD-SDI stream from the event to a centralized encoding location, which produces the compressed streams. For most readers, however, the stream or streams that you create onsite is/are the ones that your viewers will ultimately see. To create these streams, you have at least four options, as shown in Figure 11-2.

- Software
  - Telestream Wirecast
  - Adobe Flash Live Media Encoder
  - Microsoft Expression Encoder
  - Bundled with capture cards
  - Provided by live streaming service providers
  - Kulabyte XStream Live 2

- Capture/Encoder Card
  - ViewCast
  - Digital Rapids
  - Winnov

- Portable Streaming Appliance
  - Digital Rapids
  - ViewCast

- Rack Mounted Systems
  - Digital Rapids
  - Elemental
  - Inlet HD
  - Winnov

Figure 11-2. Your encoding options for live events.

The two categories on the left—software and capture/encoder card—are a bit joined at the hip. That is, all software encoders need some kind of video capture device to work; whether it's a simple DV port on your computer or notebook, a higher-end Blackmagic DeckLink card or a low-end ViewCast analog capture card.

Turning that around, all capture cards need software to configure the streams and connect to the live server. Sometimes, it's a third-party program like Adobe Flash Media Live Encoder; sometimes it's proprietary software bundled with the card, like ViewCast SimulStream, which is bundled with several high-end ViewCast cards. Sometimes, it's software provided by the live streaming service provider that you chose, like Livestream or Ustream.TV. The bottom line is, whatever software solution you choose, you'll need to find a compatible capture device, and whatever hardware device you use, you'll need compatible software.

Fortunately, in most instances, commercial software programs like Telestream Wirecast support well-defined standards, like DirectShow on Windows and QuickTime on the Mac. As you would expect, most capture/encoder cards are built to support those same standards, so the software and hardware should play together well without much effort on your part. Still, if you end up choosing your software encoder first, check the software developer's website for a list of compatible hardware devices before buying your hardware. For example, Figure 11-3 is a screen grab from the Microsoft site that contains a list of hardware products that are compatible with Expression Encoder 4.

For live encoding, the following hardware capture devices have been tested as live sources, however this list is not a recommendation or endorsement of any product:

- Black Magic Intensity
- Black Magic Intensity Pro
- ViewCast Osprey 230
- ViewCast Osprey 530
- ViewCast Osprey 450e
- ViewCast Osprey 700HD
- Winnov Videum 4000
- WinTV HVR-950

Figure 11-3. Most software vendors will present a list of known
compatible devices on their website, like this one from Microsoft's site.

This leads to the obvious question: Which should you choose first, hardware or software? In general, choose the software first for relatively simple live events with a single incoming camera or video feed and a relatively small set of outgoing streams. In contrast, for more complicated events, where you might be converting video from multiple sources and producing multiple streams in multiple formats, your first concern should be identifying a capture device that can capture and output the required streams. In that situation, buy the hardware first, and then the software.

Webcasting Software

Webcasting software runs the gamut from simple, one-source encoders like Adobe Flash Media Live Encoder to a television production studio in a single program like Telestream Wirecast, shown in Figure 11-4 from a live event I produced in 2010, story to follow.

Figure 11-4. Telestream Wirecast is a wonderfully well-featured live encoding program.

Briefly, Wirecast can mix feeds from different sources, including multiple FireWire-based cameras. Wirecast can also produce titles, broadcast disk-based images, audio and video files, as well as application screens from the same or other computers on a LAN. It's a very alluring feature set that lets you create a much more polished broadcast than do most other programs.

Here's a list of factors to consider when choosing a software encoder:

- *Can it connect and stream to your server?* Job number 1 for all software encoders is transmitting your encoded streams to your streaming server, and the webcasting programs are getting more capable in this regard. For example, while the current version of the Adobe Flash Media Live Encoder (FMLE) can only connect to Flash Servers, at NAB 2011, Adobe previewed a version that can also distribute to IOS devices. Similarly, Microsoft's Expression Encoder 4 can directly distribute to Windows Media and Silverlight clients, and can serve iOS devices using the IIS server as described in Chapter 7. Third-party programs like Wirecast are more agnostic and produce multiple formats. If you'll be using a third party live streaming service provider like Livestream or Ustream. tv, check to see if it offers a downloadable streaming encoder (both of these do), which will ensure that you can connect and transport the required stream to the service.

• *What else can it transmit?* Some programs, like Wirecast, can display graphic images, disk-based videos, application screens and other content, which is very convenient and makes for more polished presentations. Others, like Adobe Flash Media Live Encoder, only broadcast live video.

• *What about titles and transitions?* Titles, like those shown in Figure 11-4, are another useful bit of polish, as are transitions that smooth out camera switches, or the appearance or disappearance of titles.

• *Does it support multiple cameras?* Some programs, like Expression Encoder 4, support multiple cameras (and can broadcast for disk-based files) but can't produce titles and transitions.

• *Can it display slides with picture-in-picture and other mixing?* I can't speak 10 words without PowerPoint slides, so a program's ability to mix video with slides—either via picture-in-picture or some kind of 3D option like that shown in Figure 11-5—is an attractive feature. If you're broadcasting training or similar presentations, you'll probably find it valuable as well.

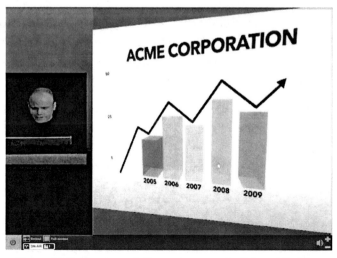

Figure 11-5. Livestream's Procaster can mix slides with video.

- **Can it capture live application screens?** If you're producing live training on how to use a software program, it's a great to be able to broadcast the program itself as part of the video feed.

Software Encoder Performance

In addition to the features-based analysis for choosing a software encoder, there's a performance aspect as well. In early 2011, I reviewed the four programs shown in Table 11-1, rating them on CPU efficiency, data rate accuracy, data rate consistency and video quality. You can read the article at bit.ly/liveencoder.

	CPU Efficiency	Data Rate Accuracy	Data Rate Consistency	Video Quality
Adobe FMLE	3	1	1	1
Kulabyte XStream 2	1	1	3	2
Microsoft EE4	4	1	2	3
Telestream Wirecast	2	1	4	3

Table 11-1. Rating the performance aspects of four software encoders, with 1 being the highest rating.

CPU efficiency rated how the software programs performed on three tested notebooks. Kulabyte XStream 2 was the most efficient (but costs more than $5,000), with Wirecast next. Both programs could produce three video streams suitable for adaptive streaming on a Core 2 Duo-based Windows or Mac notebook. In contrast, while FMLE and EE4 could produce a single stream on these systems, you needed a more powerful system to produce adaptive streams. The bottom line is that not all programs run efficiently on the same system. With the programs that have a fully functional trial (or are free, like FMLE), you should download and test performance on your capture station to make sure it can handle the load.

The next test involved data rate accuracy, where all programs produced streams very close to the target data rate. On the other hand, Wirecast and XStream had issues with data rate consistency, producing spikes in the data that could stop the broadcast if you were streaming via a limited outbound connection (See Figure 11-6).

This wasn't a show-stopper (har, har) for either product, but next time I produce an event with Wirecast over a tight DSL connection, I'll be more conservative with my data rate. For example, rather than streaming at 650 kbps over an 800 kbps connection, I'll probably drop down to 600 kbps to make sure that data spikes don't disrupt the show.

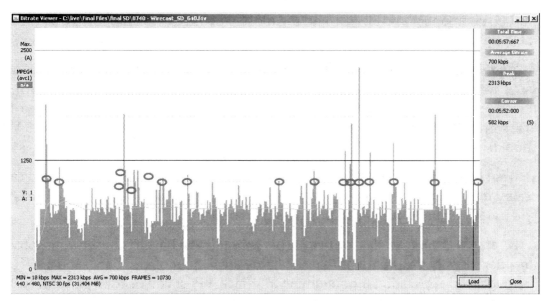

Figure 11-6. Wirecast's data rate had a few more spikes than the other encoders, and they tended to last a bit longer.

The quality ratings in Table 11-1 are overstated, as the difference between the leaders and bottom dwellers was very limited. All the programs produced sufficiently high quality for most live broadcasts, though Adobe had fewer artifacts and other issues than the rest.

Overall, to choose the best software encoder for your application, identify the candidates that provide the features that you need, and be sure to consider performance aspects as well.

That's it for software encoders; let's move on and look at the capture card/encoder side.

Choosing a Capture Card/Encoder

Here are factors to consider when buying a capture card/encoder.

- **Does it work with my software?** If you choose your software first, check the hardware compatibility list to make sure the capture card/encoder is compatible with your encoder.

- **Does it produce the required streams?** You've planned your broadcast and know how many streams you need and in what formats and configurations. Make sure the hardware you select can output those streams. Not all cards can produce HD streams, so if that's a requirement, make sure it's supported.

- **Can it connect to my video feed?** This is a big one as well. Inexpensive cards typically support either FireWire or composite/S-Video input, with more expensive cards supporting component and SD-SDI, and the most expensive cards supporting single or multiple HD-SDI inputs. Make sure the card supports the feeds you'll be working with.

- ***What incoming audio/video adjustments are enabled?*** This will typically depend upon input source, but it's nice to have brightness, contrast, hue and saturation adjustments for video, and volume adjustments for audio. Some cards that input digital video (DV, SDI) won't offer these features, while most that input (composite, component, S-Video) should.

- ***What hardware pre-processing features are supported?*** Make sure that the card can scale, crop, deinterlace and perform inverse telecine in hardware, which is more efficient and usually higher-quality than processing in software.

- ***What software is available to drive the hardware?*** If you haven't chosen your software encoder, identify the programs that the hardware vendor offers (if any) and ask if there is a published application programming interface (API) if you need to perform custom development for your live streaming application.

Once you narrow down your selections, I would sniff around the web to see how the board is faring in production environments. If the vendor has a support forum, check that, or Google "name of product" "name of company" "driver issues" or "problems." That should identify any major issues that you may be facing.

Portable Streaming Appliances

Why use an appliance rather than a notebook with streaming software? A couple of reasons. First, since they can be pre-configured back at the home office, all the onsite producers have to do is connect the camera to the device and press the streaming button.

For example, Figure 11-7 is the ViewCast Niagara 2120 streaming appliance, which I reviewed at bit.ly/niagara2120. To run an event, you turn the unit on, plug the AV feeds from the camera into connectors in the back and turn the camera on. You should see audio goosing the green LED volume meters on the bottom, and once you press the red Stream button, the V atop the button tells you that you're streaming. It's a pretty simple operational paradigm that non-technical users can easily master. Software encoders aren't rocket science, but they're more complicated than this.

In addition, unlike software encoders that vary in performance depending upon the platform they're installed upon, appliances should always perform up to their rated specifications. Even though they're usually Windows-based computers in a box—the Niagara 2120 runs Windows XP—users don't load games, shareware and other funky applications on appliances like they do their own notebooks, so the appliances should function more reliably over the long haul.

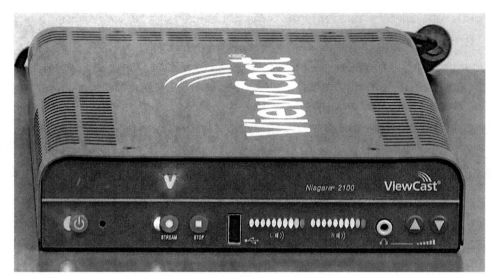

Figure 11-7. Press the record button; once the V shows up, you're streaming video.

Driving a Portable Streaming Appliance

Let's explore how these appliances operate, using the Niagara 2120 as an example. From a hardware perspective, the Niagara weighs less than 5 pounds, so it's very portable.

Again, it's a Windows XP computer in a box, driven by a 2.2 GHz Intel Core 2 Duo CPU, with 1 GB of RAM and about 110 GB of free disk space. The unit is about the size of a healthy college textbook—7.5"x11.5"x1.75"—and is cooled by a noisy fan, so you should locate the box as far from the center of action as possible. You can attach keyboard and monitor to configure the 2120 directly, or connect the unit to your LAN and configure it from any node on the network.

The 2120 accepts SD video only, up to a maximum of 480i, and produces a maximum video resolution of 720×480. Video inputs include composite, S-Video, and component, with either balanced audio input via XLR or unbalanced audio via RCA connectors.

You operate the Niagara 2120 via encoder settings that you can combine into groups to produce multiple adaptive streams. For example, a single-encoder setting might be a 640x360 stream at 500 kbps in H.264 format for Flash. An encoder group might have multiple Flash streams for adaptive encoding, or even separate Flash and Windows Media streams for multiformat presentations.

If you're sending the unit out on location with non-technical users sans monitor or keyboard, you designate the desired group of encoding parameters as "active" in the software. When your remote user presses the Stream button on the front of the unit, it executes this active group, encoding the incoming stream to the parameters set in the active group's encoder settings. If you want only a single stream, you would build a group containing just a single setting and make that the active group.

If you're encoding via the browser-based GUI (Figure 11-8), you can start any encoder or group by clicking the Streaming button on the right of the menu under "Streaming." This is how most producers will use the unit when driving it via a keyboard, mouse and monitor.

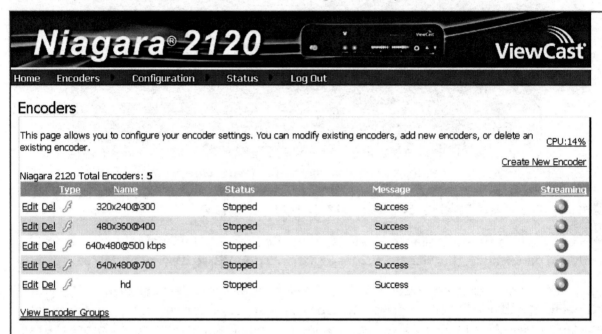

Figure 11-8. Niagara's browser-based GUI. Click the button on the right under "Streaming" to start encoding.

Once plugged into the remote network, the Niagara will acquire an IP address automatically, just like a notebook computer. If you run into problems in the field, a member of the IT staff should be able to remotely log into the Niagara via that IP address and resolve any issues, though firewalls could be an issue. Between you, me and the two dogs sleeping in my office right now, I'd bring a monitor, keyboard and mouse with me for all live events so I can run the device directly.

However you drive the unit, you create encoding settings in a screen that contains just the basics, including input, output, and audio and video bit rate (Figure 11-9). There is a much more advanced encoding options screen, though most users will likely accept the default H.264-related settings and not even glance at the advanced options.

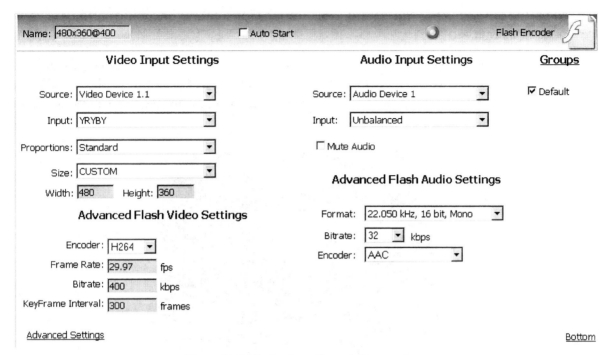

Figure 11-9. The basic audio and video settings.

Again, you assemble your encoding settings into groups, and designate one of those groups as Active. If you're operating via the controls on the Niagara device itself, clicking the Stream button encodes the incoming stream to the settings contained in the Active group.

Choosing a Portable Streaming Appliance

When choosing a streaming appliance, you should consider many of the same factors identified previously for capture/encoder cards, along with a few unique ones, which I'll cover first.

- *Touchscreen or not?* Though the units are more expensive, Digital Rapids' line of TouchStream appliances give you a full preview of streaming operation with touch-screen controls. It's nice to see that video preview, and to adjust controls directly on the unit if things go wrong. With devices without the touchscreen, you have to bring either a notebook to connect to the unit over the LAN, or a keyboard, mouse and monitor to drive the unit directly.

- *How tough is it to create my settings?* I've reviewed both ViewCast and Digital Rapids appliances (see the Digital Rapids review at **bit.ly/touchstream**), and I favor the ViewCast software because you create and implement the settings in the same program. With Digital Rapids, you use separate programs to create settings and drive the hardware, and because of some funky license restriction, you can't have both programs running at the same time, which is frustrating. The interface for creating encoding settings is very obtuse, as well.

• **How can I access it remotely?** If you'll be sending the unit out with non-technical users, find out how easily you can control the software remotely. For example, the ViewCast software is totally browser-based, so you drive it locally or remotely via the same interface, which is a nice paradigm.

Figure 11-10. Digital Rapids TouchStream appliance.

Beyond this concern, you have many of the same considerations discussed above, including:

• **Does it produce the required streams?** You've planned your broadcast and know how many streams you need and in what formats and configurations. Make sure the appliance you select can output those streams. Not all devices can produce outbound HD streams, so if that's a requirement, make sure it's supported.

• **Can it connect to my video feed?** This is a big one as well. Inexpensive appliances typically support either FireWire or composite/S-Video input, with more expensive devices supporting component and SD-SDI, and the most expensive devices supporting single or multiple HD-SDI inputs. Make sure the appliance supports the feeds you'll be working with.

• **What incoming audio/video adjustments are enabled?** This will typically depend upon input source, but it's nice to have brightness, contrast, hue and saturation adjustments for video, and volume adjustments for audio. Some appliances that input digital video (DV, SDI) won't offer these features, while most that input analog video (composite, component, S-Video) should.

- **What hardware pre-processing features are supported?** Make sure the appliance can scale, crop, deinterlace and perform inverse telecine in hardware, which is more efficient and usually higher-quality than processing in software.

Since appliances tend to be very expensive, I'd definitely check for user reviews before buying. If the vendor has a support forum, check that, or Google "name of product" "name of company" "driver issues" or "problems." That should identify any of the major issues that you may be facing.

Choosing a Rack-Mounted System

Rack-mounted systems are for high-volume applications and typically cost $20,000 and higher. While I've reviewed a system or two, I don't have a lot of experience working with encoders in this class. For an article I wrote for *StreamingMedia* Magazine in 2010 (bit.ly/webcastmasses), I asked Major League Baseball Senior VP Joe Inzerillo—who's charged with broadcasting 2,430 regular season games, plus preseason and postseason—for some tips.

For Inzerillo's work with MLB webcasting, reliability is critical. Before choosing Inlet HD's Spinnaker encoder, he held a "bake off" between multiple vendors, running them for a 5-hour pre-game/game/post-game period and analyzing factors like CPU use and dropped frames.

While an occasional dropped frame may not be a huge deal on an iPod, an increasing number of his customers are watching the 720p streaming feed on widescreen television sets, and dropped frames are very noticeable. He recommends that if you can't get a system in for simulated testing, ask the vendor to put you in touch with clients whose applications and run times are similar to yours.

Inzerillo also noted a significant disparity in the breadth and depth of H.264-related controls made available by each encoding tool. This is significant because he uses different encoding parameters depending upon factors such as whether it's a day or night game. In addition, his organization provides services to other networks during baseball's off season, and encoding parameters that work well for the center field camera in baseball may be totally inappropriate for the side-to-side action in soccer or basketball. So work through the preset creation process to make sure that you're comfortable that it provides access to all relevant encoding controls.

Beyond these considerations, Inzerillo advised to make sure that any encoder you choose can be controlled remotely, preferably with multiple encoding stations that are simultaneously viewable from a remote monitoring station. Also look for features such as scheduling, automated failover and alert notifications.

In addition to these concerns, you obviously have many of the same nuts-and-bolts issues discussed above, including:

- *Does it produce the required streams?*

- *Can it connect to your video feed?* Make sure the system can connect to your current video feeds—and, since the systems are so expensive, the feeds you plan to be using over the next two to three years.

OK, now you know what to buy to encode your streams, so let's look at your server/distribution options.

Server/Distribution Options

Once you've encoded the necessary stream or streams, you send them to the live streaming server, which distributes the streams to your viewers. This streaming server can be self-hosted, hosted in a cloud or provided by a content delivery network (CDN). Or you can host your own server and just deliver via CDN, though most smaller-scale producers are best served by addressing both the server and distribution requirements with a single service provider. A brief walk down memory lane will explain why.

In the early days of streaming media, most live event producers hosted their own streaming server. This means they installed, configured and maintained the server, built the necessary player, and produced the event. This worked for low-volume events, but most organizations didn't have the high-speed Internet connection and network infrastructure to simultaneously deliver streams to large volumes of viewers.

In part, this spawned the rise of CDNs that developed various delivery infrastructures to facilitate the efficient delivery of high-volume live and on-demand streams. Once CDNs were in place, most large-scale live event producers used live streaming servers hosted by their chosen CDN to distribute their streams. Since the CDN supplied only the server, the producer was still responsible for player creation—which was fine, since most large-event producers have programmers on staff and wanted a high degree of player customization.

As live streaming technology matured, smaller-scale producers wanted to get into the act as well, but they didn't have the programming staff necessary to create their own players and didn't really need a great deal of customization. Then, similar to the rise of online video platforms (OVP) for on-demand traffic, a new class of provider started to appear, which I call live streaming service provider (LSSP). These organizations input your live encoded stream and provide the necessary server functionality, the distribution network, and a customizable player or web page to watch the videos. Or of course, you can also embed the live streams into your own website.

For smaller-scale producers, farming out live streaming services to an LSSP saves capital expenditure and personnel cost, and you get faster access to new features because LSSPs can amortize their development costs over multiple customers. Since most LSSPs distribute their streams via CDNs, the viewing experience should be better than what you could provide from your own server and Internet connection, and distribution costs should be lower because the LSSP can aggregate streams to negotiate a better rate with the CDN.

	Host Your Own Server	CDN	LSSP
Server install, hosting and administration	You	CDN	LSSP
Player creation	You	You	LSSP
Distribution network	You	CDN	LSSP

Table 11-2. Live streaming server alternatives.

Table 11-2 shows the three alternatives we've just discussed and the division of labor for the key activities on the left. Hosting your own server is fine if you have the funds and technical staff to host, install and administrate the server and create the player, and if you'll be streaming to relatively few viewers. At the other end of the spectrum, if you're distributing high-bandwidth streams to bunches of viewers, you should consider distributing via CDN. If you're a small fry and don't have the cash necessary to install, configure and maintain a server and create a player, using an LSSP is probably your best option.

Since I think the LSSP is the best option for most readers, I'll next focus on choosing the best live streaming service provider.

Choosing an LSSP

Let's take a quick look at how the LSSP market has evolved. On one side are sites like Livestream, Ustream and Justin.tv that started as kind of YouTubes for live video, funded via advertising displayed around the live streams published on their sites. As their product offerings became more polished, these sites started offering advertising-free "white label" versions that you could rent for a monthly fee. This was a natural upgrade for successful producers who wanted to transition from the free service and still access the viewers aggregated by these sites, while selling their own advertisements to monetize the video.

At the same time, a number of OVPs like Brightcove extended their on-demand services to include live, while several companies were founded primarily to stream live events. A good example is Multicast Media, which delivered more than 50,000 live events in 2009 for more than 3,000 customers. Services from these companies tend to appeal to organizations with a well-defined group of target viewers who aren't seeking viewers from the UGC communities.

Advertising-Supported Sites with White-Label Version	OVP with Live Capabilities
Livestream	Brightcove
Ustream	Multicast Media
Justin.tv	Sorenson 360
	EyePartner

Table 11-3. The two classes of Live Streaming Service Providers.

At a high level, all of these organizations operate the same way: You point your live streams at their servers and grab an embed code to insert the live feed into your own website. As with capture devices, however, there are a number of issues you should consider before choosing an LSSP.

Cost is always a consideration, and if there's no budget for live streaming, choose a free account with Livestream, Ustream or Justin.tv. You might work through the issues presented below, however, to see exactly what you are or are not getting with the free account.

Formats Supported

Let's start with the supported output formats.

- **What formats are supported and how?** If video is truly mission-critical, odds are that you want to support both the Flash and iOS platforms, and maybe Android, Palm, BlackBerry and others. The big issue is whether the LSSP can dynamically transmux from one format to another. I talked about this back in Chapter 7, Supporting Multiple Adaptive Streaming Technologies, where I discussed how certain products and services can input one stream (or set of streams) encoded in H.264 format and dynamically re-wrap the files for viewing in another environment. This is shown in Figure 11-11 (which is also Figure 7-3), a functional diagram of the Wowza Media Server 2.

At this point, most LSSPs support iOS devices; the question is whether they can do that via a single input stream, or whether you'll have to input a discrete stream for iOS distribution. If the latter, that doubles the encoding cost and, because you'll have to create multiple streams, makes it tougher to get video out of locations with limited-bandwidth connections.

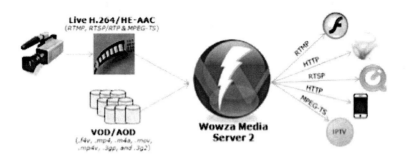

Figure 11-11. Wowza Media Server dynamically transmuxing files for viewing on different environments.

- **SD or HD?** Not all service providers stream in HD, particularly with their free products, so check availability and cost if you want HD.

- **What's the codec?** Some services use VP6 for their free services, with H.264 reserved for paying customers. H.264 is definitely the higher-quality option and the codec you want to use.

- **Are the streams adaptive or static?** If you're targeting mobile devices, adaptive streaming is critical to a good user experience. If the LSSP offers adaptive streaming, determine whether there's a limit to the number of streams you can distribute and whether the LSSP can create streams from a single incoming high-quality stream, like Sorenson Squeeze Live, or whether it simply rebroadcasts the streams that you deliver to it.

Supported Input Formats

The next area of inquiry relates to getting video into the LSSP.

- **Is there a custom capture product?** Most of the live UGC sites offer their own capture encoder—sometimes a free software encoder, sometimes a hardware/software combination that you can buy or rent to send the optimal stream configuration to the site. As with the software encoders that we discussed above, determine the features of the encoder: whether it can support multiple cameras or stream disk-based video files; whether it can handle other inputs like application screens; and whether it supports titles, bit mapped overlays, and transitions.

- **If there's no custom encoder, which encoders does the service support?** This shouldn't be a huge issue; most support any encoder that can send a live Flash stream to their server, which most software encoders can do. Still, you should ask for a list of known compatible products and select an encoder known to be compatible with the service.

- **What mobile devices does the LSSP support?** Can the LSSP stream incoming video from a mobile device? It may sound silly now, but the ability to stream from mobile devices will gain importance as devices get more powerful CPUs and more capable cameras and mobile bandwidths get more reliable.

Player Details

The player is the window through which your viewer watches your video, so features and branding are important. Ask about the following:

- **How brandable is the player and channel?** Most services offer multiple templates that you can brand and/or configure. Find out how many templates each LSSP offers, and scan through them to determine if they're imaginative or cheesy. Verify that you can add logos and other design elements to the player and channel and, if desired, make the channel and player look unique and/or visually cohesive with your own website. Make sure the player offers a competitive range of controls, such as volume adjustment and full-screen viewing.

- **How/where can you embed the stream?** Determine whether you can embed the live stream in your own website, enable others to do so and lock out certain sites. Also check that embedded player resolution is flexible or limited to fixed sizes, which can conflict with the design of your website or blog.

- **Does the player support chat and other social media features?** Interactivity is key to viewer engagement, and many LSSPs are now supporting live chat during and after your event. Most also let you promote upcoming events on Twitter, Facebook and other social media sites.

Monetization and Security Concerns

Next, determine if the LSSP supports your monetization strategy and can address your security concerns, asking:

- **Does the service support your monetization business model?** For example, if you're webcasting for lead generation, can you capture names and email addresses of your viewers? If your model is pay-per-view or subscription, does the site support that? What about links to your own advertising networks or third-party advertising networks?

- **Can you require user authentication to view the stream?** It's tough to talk about that upcoming product announcement (or any other private matter) if you can't limit who watches the stream. Find out what type of authentication the LSSP supports and other security-related features, like whether encryption is available.

Service Aspects

Next, consider the service aspects of your LSSP partner, asking:

- *What is the availability of turnkey production packages?* Many novice live event producers have limited camera equipment, AV production skills, or even IT skills. If you're trying to host a high-quality event, factors such as lighting, microphones and capture settings can make a huge difference in quality, and you'll likely want some help for your first one or two events—perhaps for all of your events. Some live streaming service providers can send a technician with all necessary equipment to your door for a turnkey webcast, but most can't.

- *What is the availability of custom development?* If you think you'll need help developing your landing page or player, check out whether the LSSP offers custom development services.

- *Can the service archive your online presentations?* Most can, but this raises a whole series of additional questions about whether the site is the best choice as your online video platform (OVP). OVPs provide a range of services for on-demand and live videos, including media management, interactivity and analytics, with varying options for player customization. Choosing an OVP based primarily upon its live event capabilities could be a tail-wagging-the-dog decision. You're better off analyzing the organization's on-demand and live capabilities as a whole.

For considerations here, check out my articles "Choosing an Online Video Platform" (**bit.ly/chooseOVP**) and "The Moving Picture: Meet the Online Video Platform" (**bit.ly/MeetOVP**).

Now you know how to choose the components necessary to produce a live web event, so let's briefly touch on a related but separate market called rich media communications.

Rich Media Communications

The rich media communications category includes presentations with PowerPoint slides and other graphics, as well as chat, polling, Q&A, and other similar features. If you're running a conference or training seminar and want your live video stream to incorporate all the elements actually used by the presenter, plus viewer interactivity, you'll need a rich media communications system.

As an example, one of the most prominent rich media communications companies is Accordent Technologies, and Figure 11-12 is a screen grab from one of its archived presentations about the company's product offerings (**bit.ly/accordentpreso**). This is an archived presentation, and you can see the various rich media components: video on the upper left, clickable links to different slides on the bottom left, slides on the right. The tabs on the lower left provide access to additional information, resources and email link. In a live presentation, there would be windows for chat and polling. Overall, it's a pretty impressive presentation for training, corporate communications and the like.

These products are very, very deep, and a comprehensive comparison would take too long. But let's identify the major players and flesh out the category a bit, so you can tell if it's interesting to you.

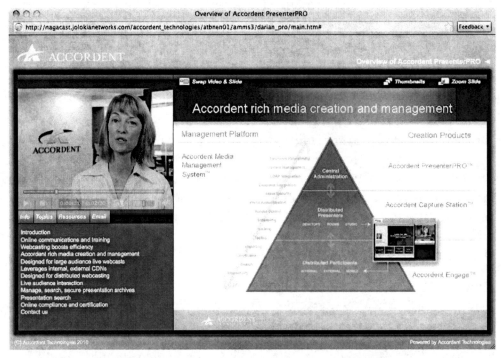

Figure 11-12. A rich media presentation courtesy Accordent.

The first questions I would ask when evaluating these products are the business model and initial capital investment. Two of the best known participants—Accordent Technologies (just acquired by Polycom) and Sonic Foundry—use the traditional software/hardware model, where you buy (or rent) and install a dedicated computer from the company, and then start broadcasting from your own website or via an arrangement with a third-party service provider such as a CDN. In addition, Sonic Foundry will host the live and/or on-demand website for you, a service that Accordent
doesn't provide.

Another vendor, Qumu, sells an installed hardware/software option, but it also offers a software-as-a-service (SaaS) option that may minimize your initial investment. MediaPlatform also offers a cloud-based service, or you can license the software and install it on your own servers. Understand the various models that are available and make sure the vendors that make your short list support the models that you're most interested in.

Next, you'll want to check which platforms the system supports, usually either Flash or Silverlight, and whether it's single-stream or adaptive. Then you'll want to identify the supported mobile players, if any.

Another critical factor to consider, particularly in a conference environment, is the amount of remote interactivity enabled or supported by the system, such as chat, polling, and Q&A. You're trying to simulate a live event with these programs, and viewers must be able to interact to feel like they're part of the program. You may also want the ability to support multiple speakers at different locations, which isn't a feature available from all service providers.

You'll also want to understand the encoding options available with each system. As mentioned, some vendors offer a dedicated hardware or software option, which may be more expensive but may also offer greater functionality and integration than third-party, off-the-shelf products.

If the webcasts are primarily for sales and marketing activities, the service will also need to support your business goals for the event—whether they're lead generation, monetization or sales conversion—plus provide tracking that lets you easily monitor the webcast's performance.

OK, that's it. I'll close a section on my first webcast, which I produced in May 2010. After reviewing multiple real-time encoding tools over the years, I was finally going live.

I've resisted the urge to reprint articles that I've published in this book, and it's probably 95% original material. But I've decided to add this one whole, since I think it captures the gestalt of your first live event. You'll also learn a bit about the nuts and bolts of a live event, as in how to point the encoder at the streaming server and the like.

My First Live Webcast

I produced my first live webcast on May 8, 2010. Like everything else, there were some huge lessons that didn't become obvious until I was there in the chair, broadcasting live. If you've got your first live webcast coming up, perhaps you'll find them useful

By way of background, this was an event sponsored by Grayson Landcare, a "community-based group of volunteers working on conservation projects that contribute to environmental, social and economic outcomes." The event was its First Annual Save Green: Money and Energy Expo, and it had a 6-hour lineup of speakers on how to build green, and how to generate power via solar, wind and other alternative energy options.

A buddy of mine was one of the directors for the event, and she asked me to shoot for DVD and YouTube distribution in the Historic 1908 Courthouse in Independence, Virginia. As long as I had a camera onsite, I figured we could stream live as well, and she agreed. We quickly checked to make sure that the courthouse had connectivity in the auditorium where the speakers would give their talks, and when we learned that it did, the deal was done.

Choosing a Webcast Provider

My first consideration was choosing the service to stream the live video. There are a number of free services like Ustream and Livestream, but these are advertising-supported and very low-touch. Though you can upgrade to a non-advertising-based service, I wasn't sure how many events Landcare would be running, so I didn't want to commit to a monthly plan.

In addition, I wanted technical support if necessary, and I was uncomfortable with the prospect of getting quick answers from a primarily UGC site. So I decided to try to run the webcast through Atlanta-based Multicast Media, a fee-based live streaming service provider. I had gotten acquainted with Multicast at some *StreamingMedia* events, and I was impressed by its product offering and client list—which includes Delta Airlines, the University of Pennsylvania and many others. I contacted the company, and it was kind enough to offer run this charity event gratis.

Choosing a Real Time Encoder

The second consideration was how to produce the stream to send to Multicast Media's streaming servers. I considered three options. First was Telestream Wirecast ($449), which gave me sexy titles and the easy ability to post informational slides between speakers (Figure 11-4). I was familiar with Wirecast, having reviewed it in the past, and this familiarity ultimately proved to be the deciding factor. I decided to run the show on my 2.2 GHz MacBook Pro, which had one FireWire port that I connected to my Canon XH A1.

Beyond Wirecast, I also considered using Adobe's Flash Media Live Encoder, which is free but doesn't offer the fancy titles and wasn't a tool that I was familiar with at that time. Thinking that I needed a backup system in the unlikely event that the MacBook Pro failed, I loaded the program on my HP 8710w notebook, which I took along with my Sony HDR-FX1 as a backup camera.

The other encoding tool I considered was the ViewCast Niagara 2120, a standalone encoding appliance that I could have pre-configured and used without a computer at the event. In many ways, the Niagara was the most advanced option, since I could have encoded and distributed multiple streams, including streams targeted for mobile devices. However, I was hoping to just get through this first webcast, not hit a home run, so I decided to use the tool I knew best.

Getting Connected

Equipment choices settled, I focused on connecting Wirecast to Multicast Media's servers. Multicast Media, which hosts more than 60,000 live events a year, makes this simple via a "Get Code" tab in its server control software. Click this, and you'll see the URLs for the primary and fallback streaming server, the stream name, and login name and password (Figure 11-13).

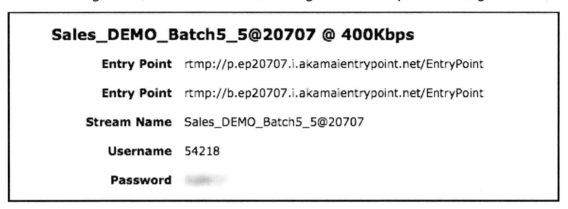

Figure 11-13. Enter this login information into your encoder, and you should be up and running.

Once you have this information, you copy and paste them into the corresponding entry fields in your encoding tool, as you can see in Figure 11-14, which shows the entry fields in the back-up Flash Media Live Encoder. Enter the user name and password, click Connect and you're up and running. Once connected, Multicast Media lets you run a test feed that is only visible in the server control software. Using this test feed, I got both Wirecast and the Flash Media Live Encoder configured and tested two days before I went live.

Figure 11-14. Here's where I entered the connection information in the Flash Media Live Encoder.

Embedding the Player

Multicast Media's Get Code page also contained the codes for embedding the live stream into Grayson Landcare's website, which took just a couple of moments. As you can see in Figure 11-15, embedding is simple; you just copy the codes and paste them into your web page's HTML editor, just like with YouTube or Vimeo videos.

Figure 11-15. The embed codes used to create the player on the Grayson Landcare website.

Choosing Encoding Parameters

After getting connected, my thoughts turned to the stream configuration I would use for the event. First, understand that in most instances, the encoding configuration set in your encoding tool (Wirecast in my case) will be streamed to your viewers; it won't be re-encoded by the service provider. You have multiple factors to consider when setting your encoding configuration, the two critical factors being the upload bandwidth from your event site and the bandwidth of your viewers.

Before the event, I measured onsite upload bandwidth at the courthouse using www.speedtest.net, and it averaged 700 kbps. This meant that my stream configuration would have to be well under this figure—otherwise, I would have trouble getting data to the server, which obviously stops the show for remote viewers.

I also knew that some viewers would watch from Australia, Indonesia and other far-flung locales. Optimally, I could have offered multiple streams via the ViewCast box, a large stream for US consumption and smaller one for overseas, but I didn't have the outgoing bandwidth for both. To be safe, I went with 320x240 resolution, 15 fps, with video at 250 kbps and audio at 32 kbps. As shown in Figure 11-16, which is Wirecast's Encoders Presets screen, I used the H.264 codec configured to the Main profile.

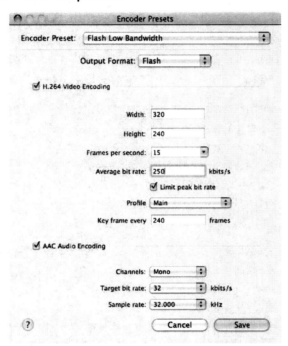

Figure 11-16. The Wirecast encoding configuration that I used for the webcast.

If you're operating in a bandwidth-constricted environment, like I was, be sure to choose constant bit rate (CBR) encoding rather than variable (VBR), because data rate spikes associated with VBR encoding can overwhelm the outgoing bandwidth and stop the outgoing stream, which can also pause the show at the receiving end (and ultimately did). In Figure 11-16, the "Limit peak bit rate" checkbox enables CBR encoding, a detail that I missed during setup, which forced me to make some fast adjustments after I went live. More on this below. As an aside, if you have a current version of Wirecast, you may notice that Telestream removed this checkbox, essentially using CBR for all encodes.

At the Show

That Saturday, I arrived at the courthouse at about 8:30 AM, an hour before the show, and first set up the XH A1 and wireless microphone. For those who care about such things, I used the AKG MP40 MicroPen microphone discreetly taped beneath the auditorium microphone, transmitting wirelessly to the AKG SR40 receiver that I connected to an XLR port on the XH A1.

Lighting (of course) would be totally au naturel—I couldn't very well fire up the tungsten lights for a green event. Fortunately, the auditorium was flanked with large windows both left and right, and Mother Nature cooperated with a sunny day that provided lots of beautiful indirect light. I set the zebras at 75, configured exposure manually and had no trouble painting the subject's faces with zebras all day long.

Next, I booted up the MacBook Pro, ran **www.speedtest.net** to confirm my earlier bandwidth findings, and ran Wirecast, which immediately found the XH A1 and connected to the Multicast Server. Then I logged into the server in a browser to check the test stream, which was also good. I opened up another browser window, and surfed to the Grayson Landcare page with the embedded video so I could monitor the live stream at the event.

Everything looked grand as the countdown clock counted backward—3, 2, 1, live. Playing the director, I cued the first speaker, but as I zoomed out of my initial wide establishing shot to a medium close up, the video I was monitoring on the Grayson Landcare website stopped dead as Ogg Theora after Google's VP8 launch. I quickly checked Wirecast's outgoing bandwidth monitor, and noticed that it was up in the low four figures, somewhere north of 1100 kbps. So much for 250 kbps video and 32 kbps audio, eh?

I flashed back to the bandwidth testing I had performed in my office, which was static camera only, and realized that the data rate was spiking with the camera motion. I checked the encoding preset, noticed the un-checked Limit Peak Bit Rate checkbox, cursed myself for not seeing it beforehand, and clicked the checkbox. The data rate had already stabilized with the static camera, so I couldn't tell if I had fixed the problem or not.

Next time I zoomed the camera out, I learned the bad news—the data rate spiked again and the video stopped again on the Grayson Landcare site. I should say that this wasn't a total calamity—the video simply started playing again a few moments later with no content lost, but the stoppage was frustratingly avoidable. I maintained a stable medium shot until the first speaker finished, then stopped the broadcast, unchecked and checked the Limit Peak Bit Rate checkbox, saved the preset, restarted the broadcast, zoomed around to see if the problem persisted, and found it was solved.

Again, when you're preparing for your first live event, check outgoing bandwidth onsite, make sure your outgoing stream is comfortably less than the outgoing bandwidth, and make sure you're producing a CBR stream.

What else would I do differently? Well, when the last speaker stood up in a Madras shirt, I saw the data rate start to strain against the upper limit and quality drop noticeably, and I remembered that I should have advised the speakers to wear conservative, solid colors. I also realized that I had gotten lucky with the background, which was a solid beige that contrasted well with the clothing worn by all the speakers. Had the background been wallpaper with a herringbone pattern or other fine detail, the quality of the entire webcast could have been compromised.

As part of your live event prep, determine what the background is going to be and make sure it's compressible. If you'd like a guide to choosing a compressible background and clothing, download the PDF from my 2010 StreamingMedia East presentation "Video Production for Streaming" at **bit.ly/streamingproduction**, which covers both topics in detail. I just forgot to apply the lessons to my own first webcast.

Finally, if you haven't used a program like Wirecast before, create a short checklist for stopping and starting each segment—particularly if you have multiple speakers rather than one long show. This is especially true if you're running multiple pieces of gear, like recording tape in the XH A1 and driving the webcast, as well as working sound and lighting. I made several small omissions early on, like forgetting to press record on the camera, and these mistakes stopped once I created and observed the checklist.

So that's my first live webcast. Though all I remember are the mistakes, the ratings apparently were high from LA to Australia to Sri Lanka—no doubt, in part, because of the big outgoing pipes of Multicast Media. In fact, my buddy has asked if I want to go to Sri Lanka next year to observe Landcare in action overseas. I've already suggested that we produce a documentary of the event, which of course I'll write about in an article titled "My First Documentary." I'll be sure to read all of my own training manuals before I start shooting that one, though, I promise.

Conclusion

That's it for live; next chapter, you'll learn how to accelerate your encoding of on-demand files.

Chapter 12: Accelerating Encoding on Multi-Core Workstations

No matter how fast your workstation, the one constant in video encoding is that it never occurs as fast as we'd like. Fortunately, if you have a multiple-core workstation, there are some steps you can take to accelerate encoding. Specifically, there are two:

- If you're producing with Apple Compressor, you can enable Qmaster, which opens up multiple compression interfaces to encode files more efficiently.

- You can open multiple instances of some encoding programs and encode in each instance separately.

In this chapter, you'll learn how to enable and use Qmaster on a single workstation and in LAN-based clusters. You'll also learn which programs can open multiple instances, when it makes sense to try encoding with multiple instances and how to do it. I'll conclude with a brief look at H.264 hardware acceleration products.

Working With Qmaster

Qmaster is an application Apple created to allow Compressor and other rendering-intensive applications to use other Macs on a network cluster to share encoding tasks. However, if you have a multiple-core system, you can also use Qmaster to create a cluster from the multiple cores to accelerate rendering in Compressor. It's a two-step process: first you have to configure and start Qmaster, and then you have to choose the new cluster for encoding in Compressor.

Start in the Mac System Preferences window, and click Apple Qmaster to open the Qmaster configuration window (Figure 12-1). If you get a message that says "To use the 'Apple Qmaster' preferences pane, System Preference must quit and reopen," click OK and you'll see the screen shown in Figure 12-1.

In the Apple Qmaster window, choose the QuickCluster With Services radio button. Note that this is critical, as neither other option will allow you to access the cluster from within Compressor. Then click Compressor—Distributed Processing for Compressor in the Services window.

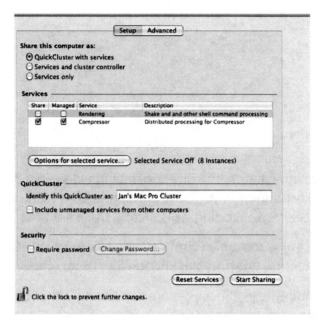

Figure 12-1. Configuring Apple Qmaster.

If you messed up at this stage, the Identify This QuickCluster As text entry box remains grayed. This is the next step, and if you can't insert a name here, you've clicked the wrong sharing option. When you get it right, name your QuickCluster something memorable—preferably machine-specific so you can easily identify it later—and click the Options for Selected Service button to open the window shown in Figure 12-2.

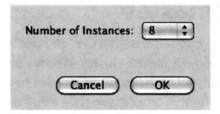

Figure 12-2. Choose how many cores to use during rendering with Compressor.

This is where you choose the number of cores available for Qmaster. If you want your Mac totally devoted to rendering, choose all available cores. On the other hand, if you need to perform other tasks while rendering, select a lesser number of cores to retain cycles for other activities. Then click OK to close this window.

Then, click Start Sharing on the bottom right of the Qmaster configuration screen shown in Figure 12-1 and close the Qmaster configuration window. Next time you run Compressor, when you submit the encoding task, be sure to send it to the cluster you just created, not to This Computer (Figure 12-3).

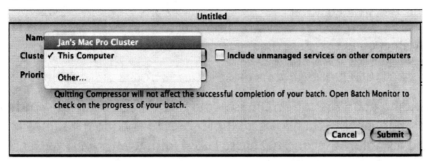

Figure 12-3. Make sure you assign the encoding task to the new cluster, not This Computer.

This will run the encoding job on the cluster, and if you check Activity Monitor while you're rendering, you should see multiple CompressorTranscoderX instances as in Figure 12-4.

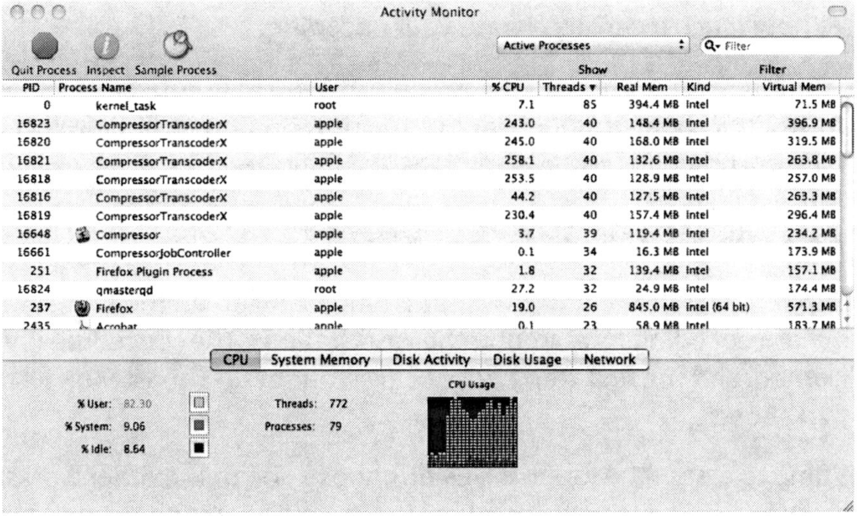

Figure 12-4. Compressor is encoding multiple files simultaneously, which is much more efficient than one at a time.

Table 12-1 presents the results of some tests that I ran while writing this chapter. In both tests, I encoded eight 1-minute DV files, first serially without Qmaster, the second time in parallel with Qmaster. As you can see, your results will vary by codec; the Apple H.264 codec appears to work more efficiently with multiple cores than x264, so the time saving is greater. Still, the

25% time savings achieved with the x264 Encoder is nothing to sneeze at if you're in a hurry to get home or deliver work to a client. For the record, I performed these tests on a 2 x 2.93 GHz quad-core Intel Xeon Mac Pro running Mac OS X 10.6.6.

	Apple H.264 Codec	x264 Encoder
Encode serially	7:47	3:37
Encode with Qmaster	2:53	2:43
Reduction in encoding time	63%	25%

Table 12-1. Reduction in encoding time from Qmaster (min:sec).

Note that depending upon the codec, Qmaster may accelerate single-file rendering as well as multiple-file rendering, since it allows multiple cores to encode the same file. For example, when I produced a single 90-second HD file to H.264 format without Qmaster, it took 7:08 (min:sec). With Qmaster, it took 4:28, a time savings of about 37%.

Overall, when you're producing multiple files, enabling Qmaster should work for any codec supported by Compressor—including MPEG-2, MP3 and others—though the time savings will definitely vary by codec. If you're encoding single or multiple files, set it up, run tests with and without, and see how much it can accelerate your encoding.

One final note: Qmaster works only when you're outputting in Compressor, not when you output via Final Cut Pro's Share option. Whether you're encoding an FCP project or standalone file, you have to be encoding in Compressor to access Qmaster.

Qmaster and Multiple Workstations

If you have multiple workstations with Final Cut Studio installed on the same LAN, you can use Qmaster to share encoding jobs among the workstations. The procedure is very similar to what you just learned, and the first step is identical—you set up Qmaster on all the workstations as explained above.

Then, when it's time to assign the job to a cluster, choose one of the other workstations that should be available in the Cluster selection box. In Figure 12-5, I'm choosing Jan's Other Mac Workstation as my target, not This Computer, which encodes without Qmaster, or Jan's Mac Pro Cluster, which is the multiple-core cluster I set up earlier. Then click Submit as normal.

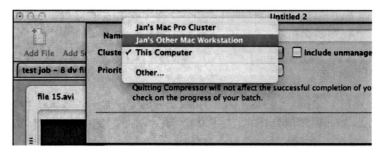

Figure 12-5. Assigning the job to another computer on the cluster.

Then, if you open Batch Monitor, you can follow the progress of the encoding. First, Compressor copies the necessary source files over to the other computer (Figure 12-6).

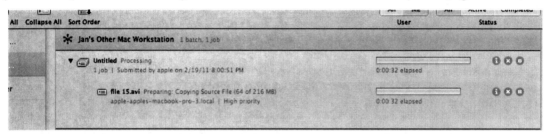

Figure 12-6. Copying the source files over to the other workstation.

Once encoding starts, you can monitor the progress of the encode just like any other project, as you can see in Figure 12-7.

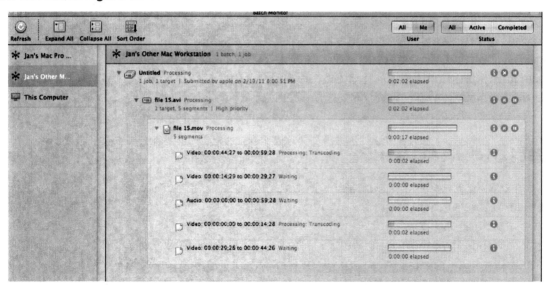

Figure 12-7. Encoding on the other workstation.

Once the encoding job is complete, Compressor transfers the encoded file to the location specified in the batch window and deletes the source files from the remote computer. These procedures should get you started, but if you need more information on how to set up a Compressor cluster via Qmaster, check out **bit.ly/qmastersetup**.

Qmaster Caveats

A couple of notes from the field. Last year, I helped a large corporation fine-tune its streaming encoding settings. While enabling Qmaster on one of its editing stations, a Mac technician that wandered by had some disparaging descriptions of Qmaster that I can't repeat in this G-rated book. He wouldn't get specific, and I had been running Qmaster for several years without problem, but I figured the most likely issue would be instability, so I Googled "Compressor Qmaster" and "unstable" and got 210 hits.

One of them was this tech note from Apple:

> However, with large, complicated jobs, such as HD H.264 transcodes, your computer may not have enough memory to support multiple services running at the same time. In this case, your computer may become unstable. Be sure your computer has an adequate amount of RAM before configuring it to support running multiple services. A minimum of 1 GB of RAM per service is recommended (**bit.ly/qmaster**).

You should definitely take this into consideration when selecting the number of instances to run simultaneously via Qmaster.

I also found a post on the Apple discussion boards—"Why is Qmaster so unreliable" (**bit.ly/qmasterreliability**)—and several other negative posts on the *Creative Cow* boards. With the number of licensed Final Cut Pro users now more than 1 million, it's not surprising that these types of problems have cropped up, but I'm guessing they're idiosyncratic, rather than systemic. As I've said, Qmaster has worked well for me, but check it out on test projects before you take it into production, and if you experience unstable operation, return to the Qmaster configuration screen and simply click Stop Sharing.

Multiple Instances

When a program encodes files serially—or one file at a time—CPU use is typically not that efficient, and throwing multiple cores at the problem won't accelerate encoding. Fortunately, with some programs, like Microsoft Expression Encoder and Sorenson Squeeze, you can open multiple instances of the program and create your own ad hoc parallel encoding system. You can see this in Figure 12-8, where I have four instances of Expression Encoder running on my 12-core (24 with hyper-threading) HP Z800, a beast of a multiple-core system if there ever was one.

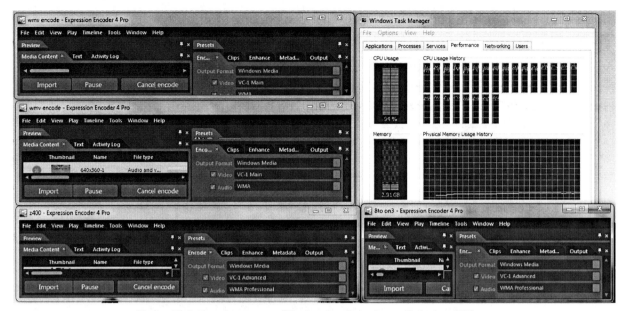

Figure 12-8. Four instances of Expression Encoder really boost CPU use.

With a single instance of Expression Encoder running and encoding a file to H.264 format, CPU use, as measured by Windows Task Manager, is less than 25%. However, if you load multiple instances of the program and start encoding in all of them, you can boost utilization to the 94% shown in the figure, or even higher. By the way, to open Windows Task Manager in Windows Vista/7, click Ctrl-Alt-Del, then choose Start Task Manager (or right click on the Taskbar and choose Start Task Manager). Click the Performance tab in Task Manager to view CPU use.

Typically, you only want to open multiple instances when:

• You're producing multiple files—this approach won't help when encoding a single long file.

• The encoding program encodes serially, not in parallel. When Squeeze 7 implemented parallel encoding of files to the H.264 and WebM formats, much of the utility of this technique was gone. However, if you still have Squeeze 6.5 or lower, it can be a real time saver. In addition, if you're producing Windows Media files with Squeeze on either the Mac or Windows platforms, using multiple instances can still accelerate encoding. If you're still using that oldie-but-goodie Flix Pro to encode VP6 files, you can load multiple instances on either the Mac or Windows platforms and dramatically reduce encoding times on a multiple-core workstation.

Unfortunately, not all programs support multiple instances. Specifically, neither Adobe Media Encoder nor Telestream Episode supports multiple instances on either Mac or Windows. No biggie if you have Episode Engine, because that can encode multiple files in parallel, but Episode produces serially, and Episode Pro can only encode two files simultaneously.

As you can see in Table 12-2, when multiple instances are available, they can significantly reduce encoding times. In these tests, I encoded eight 1-minute DV files twice: the first time encoding serially in one instance of the program, the next time opening eight instances.

	Squeeze Windows	Expression Encoder 4	Squeeze Mac
Codec	Windows Media	H.264	Windows Media
Encode serially	24:08	12:23	29:55
Encode with eight instances	7:42	6:33	18:40
Reduction in encoding time	68%	48%	38%

Table 12-2. Reduction in encoding time by encoding with multiple instances (min:sec).

Time savings will vary according to file size, codec, encoding tool and the number of instances, but if you have multiple files to encode, it's almost always worth a try.

Opening Multiple Instances: Expression Encoder

With Expression Encoder, you open multiple instances by choosing the program via the Start menu repeatedly, or, if you have a desktop or taskbar icon, clicking that repeatedly. Each separate instance is a complete functioning version of the program, and I've never experienced any kind of conflict or contention when encoding with multiple instances. Load your files, choose your presets and click Encode, then move to the next instance.

Opening Multiple Instances: Squeeze Windows

With Sorenson Squeeze for Windows, the procedure for opening multiple instances is identical to that of Expression Encoder: You choose the program via the Start menu repeatedly, or, if you have a desktop or task bar icon, clicking that repeatedly. Each separate instance is a complete functioning version of the program, and I've never experienced any kind of conflict or contention when encoding with multiple instances. Load your files, choose your presets and click Encode, then move to the next instance.

Opening Multiple Instances: Squeeze Mac

With Squeeze for the Mac, you have to jump through one additional hoop to get multiple instances going. Specifically, you have to create separate folders in the Application folder for each instance of Squeeze. You don't have to reinstall the program multiple times; you just need to create a folder and copy in the Squeeze application.

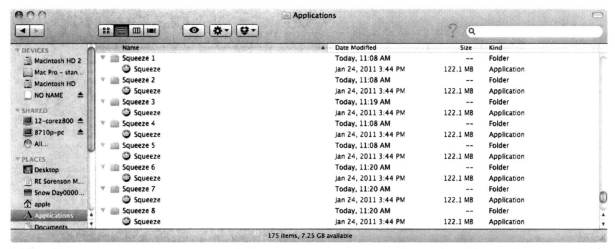

Figure 12-9. Multiple Squeeze folders containing the Squeeze application.

In Figure 12-9, you can see what the end result should look like if you want to be able to run up to eight instances of Squeeze on the Mac. Typically, what I do is create a single folder with Squeeze in it (Squeeze 1) then copy and paste that folder seven more times. Then I rename each folder in order for neatness, though obviously that's not required. If you want to run the separate programs from the Dock, you'll have to create a separate icon on the Dock for each instance.

Once you're done, click each separate Squeeze application to run it. Figure 12-8 shows a screen shot of eight instances of Squeeze running on my 8-core Mac Pro, pushing overall utilization to more than 80%. Again, you wouldn't do this for H.264 encoding with version 7 because Squeeze encodes H.264 in parallel, but if you have Squeeze 6.5 or earlier or are producing Windows Media files, give it a try.

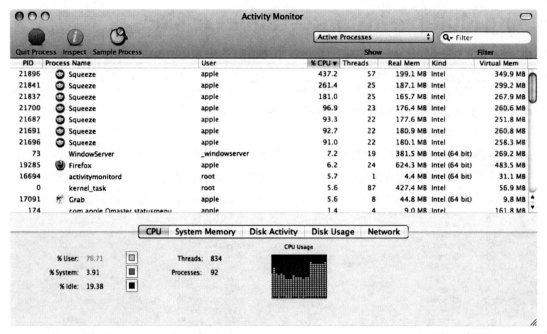

PID	Process Name	User	% CPU ▼	Threads	Real Mem	Kind	Virtual Mem
21896	Squeeze	apple	437.2	57	199.1 MB	Intel	349.9 MB
21841	Squeeze	apple	261.4	25	187.1 MB	Intel	299.2 MB
21837	Squeeze	apple	181.0	25	165.7 MB	Intel	267.9 MB
21700	Squeeze	apple	96.9	23	176.4 MB	Intel	260.6 MB
21687	Squeeze	apple	93.3	22	177.6 MB	Intel	251.8 MB
21691	Squeeze	apple	92.7	22	180.9 MB	Intel	260.8 MB
21696	Squeeze	apple	91.0	22	180.1 MB	Intel	258.3 MB
73	WindowServer	_windowserver	7.2	19	381.5 MB	Intel (64 bit)	269.2 MB
19285	Firefox	apple	6.2	24	624.3 MB	Intel (64 bit)	483.5 MB
16694	activitymonitord	root	5.7	1	4.4 MB	Intel (64 bit)	31.1 MB
0	kernel_task	root	5.6	87	427.4 MB	Intel	56.9 MB
17091	Grab	apple	5.6	8	44.8 MB	Intel (64 bit)	9.8 MB
174	com.apple.Qmaster.statusmenu	apple	1.4	4	9.0 MB	Intel	161.8 MB

CPU   System Memory   Disk Activity   Disk Usage   Network

% User: 76.71   Threads: 834   CPU Usage
% System: 3.91   Processes: 92
% Idle: 19.38

Figure 12-10. Here are the eight instances of Squeeze encoding away in Activity Monitor.

By the way, if you want to gauge CPU use on the Mac and open Activity Monitor, you can find the applet in the Utilities folder in your Mac's Applications folder. I use it so much that I have it parked in my Dock for easy access.

Hardware Acceleration

I've reviewed the Matrox CompressHD product for both the Mac and Windows platforms, though the reviews are a bit long in the tooth. For the Mac, you can find the review at bit.ly/compresshdmac.

Briefly, I found that the hardware unit did encode significantly faster than Compressor, and that output quality was better than the Apple codec. I did not test against the faster, and much higher-quality, x264 encoder, though I'm guessing that x264 would output higher quality than Matrox and would reduce the time savings significantly. The bottom line is that if you're looking to cut your Mac encoding times, try x264 first, then try Matrox.

You can find the Windows review at bit.ly/compresshdwindows. Here's what I concluded:

> To summarize my findings, on Windows, CompressHD is best suited for rendering HD input to high-resolution, high-data-rate formats such as Blu-ray or 720p files to upload to YouTube. Files produced at this resolution were equivalent in quality to those produced by Adobe Media Encoder, and they generally encoded significantly faster. In addition, I was able to load the Blu-ray files that I produced into Adobe Encore and compile a Blu-ray Disc with no problem.

In contrast, I would avoid the Windows version of CompressHD for producing H.264 files for streaming; CompressHD couldn't produce files that met the target data rate of my standard 500 kbps H.264 file, which Adobe Media Encoder easily could. Even at the higher data rate, the low-bit-rate files produced by CompressHD were substantially lower in quality than those produced by Adobe Media Encoder. In addition, with a minimum audio data rate of 128 kbps, CompressHD is not well tuned for streaming, since 64 kbps, or even 32 kbps, can often suffice for speech-only files.

I'm sure that Matrox has cured some of these ills, but Adobe Media Encoder is both faster than Compressor and produces higher-quality output, setting a much higher bar for a hardware coprocessor.

In addition, though MainConcept's NVIDIA-based GPU acceleration technology as implemented by Microsoft and Sorenson has been disappointing initially, at some point the companies should get it right. Once that's available, the potential use cases for standalone H.264 co-processing cards to accelerate streaming output gets much, much smaller.

Finally, though I doubt few readers need to accelerate Windows Media encoding, here's a link to a review that I wrote on the LSI Tarari Encoder Accelerator LCPX-6140 for Windows Media (bit.ly/tarari).

Conclusion

Almost home! Now that you know how to create multiple files quickly, in the next chapter, I'll show you several streaming file analysis tools that will help you check your work.

Chapter 13: File Analysis Tools

For the most part, streaming players provide a pitiful amount of usable data, which makes programs that provide insight into the content of these files invaluable to compressionists. In this chapter, I'll introduce you to the tools that I use daily, starting with the tool that's installed on all of my computers, Jerome Martinez's MediaInfo. You can see a video describing two the tools, MediaInfo and Bitrate Viewer, at bit.ly/twovidanalysistools.

MediaInfo

MediaInfo is a cross-platform tool that offers an extensive and often unique range of data, as well as the ability to export file-based information for printing or further analysis. It also supports pretty much every codec that I've ever tried to load, including Ogg Theora and WebM, plus the normal H.264, VP6, WMV, MOV, MPEG-1 and MPEG-2. And MediaInfo details the bits per pixel for the file, which is the single most important compression-related metric (see Understanding Bits per Pixel, in Chapter 5, for more information).

The Windows version is available in 23 different languages, including simple and traditional Chinese, and supplies more file-related data, can open multiple instances and offers more data views than the Mac. You can download both versions at mediainfo.sourceforge.net/en. Both are free, though donations are gladly appreciated.

Both versions load files using drag and drop or via traditional menu or button controls. In the Windows version, you have six different views, including text, HTML, and the Mac-style tree view. You can set which view opens by default by clicking Options > Preferences (I like the Tree view), and click the Explorer Extension to make MediaInfo appear in the right-click menu when you click a file in Windows Explorer (Figure 13-1). Very handy.

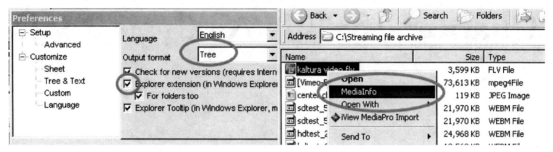

Figure 13-1. Selecting the Tree view and making MediaInfo accessible from the right-click menu.

Click Debug > Advanced Mode in the Windows version, and the program shows about three times the data, though most of the critical data is available in the Basic view.

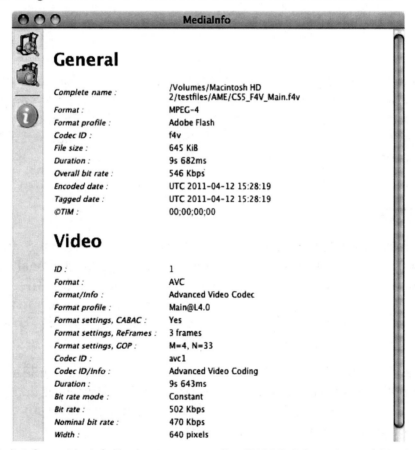

Figure 13-2. MediaInfo provides info like the entropy encoding (CABAC), B-frame interval (M=4), I-frame interval (N=33) and Bits/(Pixel*Frame), which I call bits per pixel.

The Mac version of MediaInfo can only export a simple text file, while the Windows version can output CSV, HMTL, text, and custom formats. The Windows version can also analyze multiple files simultaneously, either displaying all results in a single instance of the program, or exporting a consolidated report.

For all files (Figure 13-2), both MediaInfo versions show the resolution, data rate and other basic stats relating to both the audio and video components, including whether they were encoded with constant or variable bit rate encoding. For H.264 files, you can see the profile used, whether the file was encoded with CABAC or CAVLC entropy encoding, the number of reference frames, as well as the B-frame interval (M=4, which means a B-frame interval of 3), I-frame interval (N=33) and Bits/(Pixel*Frame), which I call bits per pixel.

For WMV files, you get the same basics, but MediaInfo doesn't identify multiple-bit-rate files, or which Windows Media SDK was used to produce the file; for that, you'll need WMSnoop detailed below. Both Mac and Windows versions can analyze VP6 files, identifying codec, resolution and similar parameters, but little else. MediaInfo also provides no information about whether the file is hinted for streaming or whether it was produced with the Fast Start option.

Though the Mac version is less full-featured than the Windows version, it's the only option I found for analyzing a broad spectrum of files on the Mac, making it a natural for most producers. The tool also reveals enough unique file characteristics on Windows—like VBR/CBR for Windows Media files, and profile and CABAC/CAVLC for H.264—to make it invaluable for most Windows producers. Let's put it this way: It's the only tool I have on all my computers, from netbook to high-end workstations, Mac and Windows.

Bitrate Viewer

Bitrate Viewer is a Windows-only tool that you can download from www.winhoros.de/docs/bitrate-viewer though I'm not sure how much longer the tool will be free. You can see the tool in all its glory in Figure 13-3.

The individual spikes represent the data rate of the file at each 1-second interval, while the wavy faint blue line is the average data rate of the file. On the upper right, you can see the average bit rate, which is always handy, plus the peak bit rate in the file. On the lower left, you can see resolution and frame rate statistics.

I like Bitrate Viewer because it instantly identifies issues that may cause playback problems. For example, in the file shown in Figure 13-3, about 80% of the way through the file, you see a large area that extends above the average bit rate line. This is a data spike that could interrupt playback.

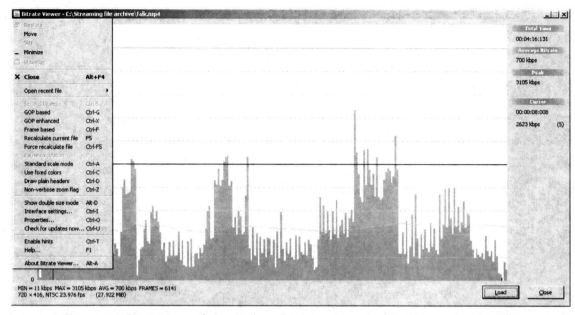

Figure 13-3. Bitrate Viewer shows you how the data rate varies over the duration of the file.

You'd have to guess that this file was encoded with variable bit rate encoding without a constraint, or with a constraint that was too high. If users reported interruptions during playback, you would know exactly why, and exactly how to fix it (either encode using CBR or tightly constrained VBR.

One negative about Bitrate Viewer is that in its default mode, the window is too small to be useful. You can double the size by clicking the icon in the upper left and choosing Show Double Size Mode, but the larger window displays an annoying message about how that feature will only be available in an upcoming for-fee version (Figure 13-4).

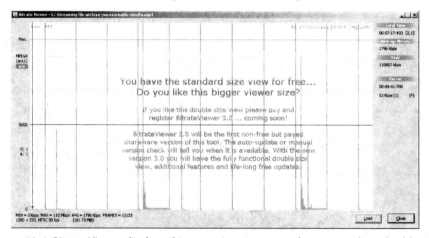

Figure 13-4. Bitrate Viewer displays this annoying message when you scale to double size.

The workaround I found quite by accident was to open multiple instances of the program, double the size of several of them (making the text appear) then returning to the menu and choosing Show in Standard Size. Usually (but not always), this makes the text disappear in the remaining instances that are still double-sized.

I'm not sure this will still work that way when you download the program, but until then, it's a great little utility that provides information you can't get anywhere else for free. Even at $10-$20 per copy, this program would be easily worth it.

GSpot

GSpot is a free, Windows-only file tool you can download from **gspot.headbands.com**. To run the program, you download a zipped file containing GSpot.exe, which never installs. Instead, you just click the EXE file when you want to analyze a file, which runs the program. I prefer not to install shareware programs, and I like this mode of operation. Once the program is up and running, you can load files via a File > Open menu command, or via drag and drop.

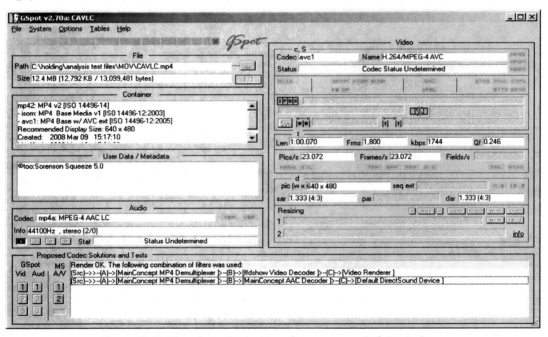

Figure 13-5. GSpot has a logical interface composed of six windows.

GSpot was originally designed as kind of a geeky diagnostic tool to help identify the codecs necessary to view a particular file. As such, it's got some compelling features, such as the ability to display all codecs and filters on your computer and track the video and audio rendering chain that displays your media. The program also offers some great functionality for MPEG program or elementary streams, where it can show groups of pictures color coded by frame type, with overlays for B-frame redundancies and actual frame numbers in the GOP. Definitely fun stuff.

However, most of the Video section, including these details, remains grayed out for Windows Media files, as well as MOV and FLV files, so it's not quite as useful for streaming files. There are some bright spots, such as revealing the date of file creation as well as any metadata packed with most files. You can see the latter in Figure 13-5, where Sorenson Squeeze 5.0 sneaks in a mention as the encoding tool. GSpot also calculates the frame quality (Qf), which we've been calling the bits per pixel.

For all streaming formats, you get the basics, with video data rate and frame rate calculated rather than simply reported from the file header or metadata. Strangely, the program failed to display a frame rate for all Windows Media files that I tested, though it worked for other streaming formats.

When analyzing Windows Media files, GSpot fails to reveal any details of the multiple streams, or the Windows Media SDK version used to produce the file. Similarly, with Flash, GSpot provides basics that may not be available in your Flash Player, such as codec and calculated frame rate and data rate, but no information you can't get from MediaInfo.

With H.264 files, GSpot again falls behind MediaInfo, failing to provide profile and level, or to show whether the file was produced with CABAC or CAVLC encoding. GSpot also doesn't show the audio data rate for all AVC-encoded files, which MediaInfo always provides.

You can configure GSpot to write out reports containing all reported data for each file analyzed, a simple way to track and accumulate file-based data. You can create separate reports for each file or accumulate all reports in a single file and analyze a folder full of files in batch, another nice convenience. Overall, GSpot is competent, but its primary strengths lie outside the streaming market, and it's best suited for MPEG-2 and AVI file analysis.

FLV Player

FLV Player is the first utility that I download to any new Windows machine because it plays FLV files of any format (SV3, VP6, H.264). Even better, when you click the info button on the lower left, an info window identifies the audio and video codecs, file resolution and frame rate, and total reported (rather than calculated actual) audio and total data rates. You can download the utility at **mdvisser.nl/flvplayer**.

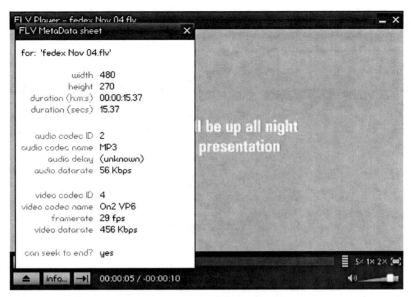

Figure 13-6. Martijn de Visser's free FLV Player gives you lots of good information right off the bat.

Be careful when you install the program, because sometimes it comes with shovelware like the Yahoo Toolbar. There is no malicious spyware, but to make a few well-deserved bucks, apparently Martijn has hooked up with several companies seeking broader distribution. Pay attention during the install; just say no at the right time, and you'll be fine.

Inlet Semaphore QC

Inlet Semaphore QC ($995) is the only true quality-control tool of the bunch, though there are a number of issues that affect utility and usability, and updates are few and far between. Make no mistake, however; if automated multi-codec quality control is your goal, the Windows-only Semaphore is the only product that fits the bill. Don't worry; you don't need to spend $995 to test out Semaphore, as there's a free trial available. Check it out at www.inlethd.com/?q=products/semaphore.

Though the two roles are obviously related, you can use Semaphore as a quality-control engine and as a file viewer with extraordinary visual analysis and file details. Let's start with the first role. At Semaphore QC's heart is a series of configurable alerts, each specific to a quality-related issue.

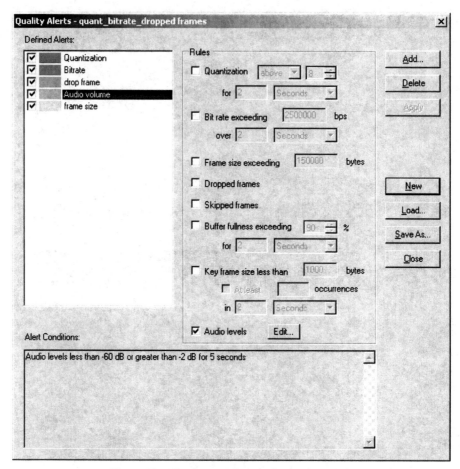

Figure 13-7. Setting up alerts in Semaphore QC.

For example, In Figure 13-7, I've configured alerts for five potential problems. Unique to Semaphore is quantization level, which measures the compression required to produce a frame, ranging from 1 (lowest compression) to 31 for Windows Media and MPEG-2, and to 51 for H.264 and VP6. While the effects of excessive quantization vary from file to file, once levels exceed 8 (for Windows Media and MPEG-2) or 25 (for AVC and Flash), visual artifacts may start to appear.

Other alerts are less exotic but equally useful. I've configured alerts to let me know if the file bit rate exceeds a certain level for a particular length of time, if there are any dropped frames, if audio volume exceeds or falls beneath certain levels, and if any particular frame size exceeds a specified number.

Once I've set up these alerts, I can view the file in Semaphore, and the color codes will alert me to problem areas in the file in the timeline just below the video window. Though you may not be able to see the colors in the screen shot, a red bar means that quantization levels exceed 8, and the blocks in the image show that excessive quantization was a valid predictor of quality concerns. The green bar indicates that the file bit rate exceeded certain levels, while the faint orange line indicates a dropped frame.

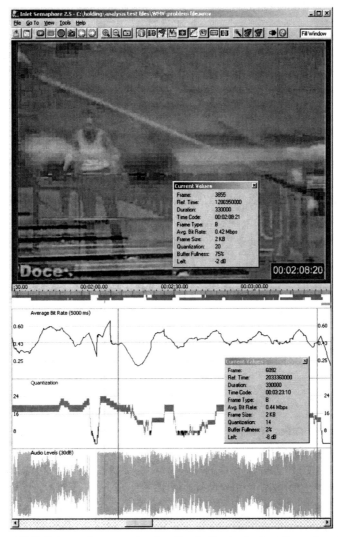

Figure 13-8. Color codes atop the timeline indicate alerts at those locations.

Semaphore can also produce HTML reports showing the alerts in summary form or each location of a problem frame. I can also run this analysis in batch mode, throwing multiple files (or folders full of files) at the Semaphore analysis engine. The program will analyze each file and produce a report for each one. You'd need to open each HTML file to see if there were problems; missing and sorely needed is a summary report that can identify all files with problems in the batch. Still, you have to walk before you can run, and Semaphore is miles ahead of any competitor in terms of automated quality control.

That's the quality-control angle; now let's look at Semaphore as a file viewer. You can open up multiple instances of Semaphore and load files via drag and drop. In addition to the visual graphs shown in Figure 13-8, Semaphore can also display a File Properties page with details that vary by format, as discussed below.

The program has a highly capable integrated file viewer. Most notably, you can move through the file frame by frame, with the current values for each frame shown in the eponymous window. The viewer identifies whether a frame is I-, B-, or P-frame, so you can determine the number of B-frames per GOP or whether your encoding tool is inserting key frames at scene changes.

Beneath the timeline are multiple graphs that you can configure to show quantization level, bit rate and audio volume. You can also examine buffer fullness and frame size, and analyze two files at one time. For example, in Figure 13-9, I'm analyzing two files: one produced using VBR, which is jumping all over the place, the other CBR and hugging the midline. Which file would you rather stream to a mobile customer?

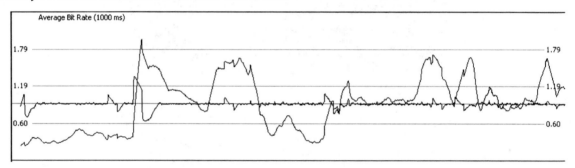

Figure 13-9. Comparing the bit rate of two files: one produced in VBR, the other CBR. Can you tell which is which?

For Windows Media Video files, Semaphore shows whether the file was produced with VBR or CBR techniques, but it listed every file as CBR, even though Semaphore later contradicted itself and reported that the file was encoded via VBR. Sounds like a bug to me—I'd trust MediaInfo for this information more than Semaphore. Though Semaphore identified multiple-bit-rate files, details about each stream were scant; you'll have to check with WMSnoop for resolution, data rate and the like.

Semaphore won't load Flash files produced with the Spark codec, an understandable decision on Inlet's part not to devote the resources necessary to analyze what is essentially a dying codec. Still, a clearer error message might avoid some confusion as to why "this file can't be rendered."

The most serious limitations relate to QuickTime files, where Semaphore can't load files that have been hinted for streaming or have a compressed header for Fast Start. So if you're producing files for the QuickTime server, or for progressive download, Semaphore is definitely not for you. On a positive note, H.264 producers can determine B-frame statistics by paging through a couple of GOPs and identifying the particular frame types, though you'll need to resort to MediaInfo to figure out whether a file was encoded with CABAC or CAVLC.

Overall, Semaphore is comprehensive and highly usable—though, as mentioned, it's Windows-only. Beyond the QuickTime-related limitations, it's hard to imagine a serious streaming producer who couldn't benefit from having the product in his or her shop.

QuickTime Player 7

QuickTime Player 7 can open multiple instances on Mac and Windows, which makes it an essential playback tool for many producers. QuickTime comes in two versions, Player and Pro, with Pro offering more encoding and diagnostic features. On the Mac, there's also QuickTime X, which came with Snow Leopard, though it has much fewer diagnostic features than Version 7.

With the Player versions of QuickTime 7 (Windows and Mac) and both versions of QuickTime X, you can load a file, choose Window > Show Movie Inspector, and get the information shown in Figure 13-10. Nowhere near as much as MediaInfo, but helpful, nonetheless.

Figure 13-10. Information available in all versions of QuickTime, Mac and Windows.

If you upgrade to the Pro version of QuickTime Pro 7 ($29.99), you can access unique data relating to hinted streaming files. Choose Window > Show Movie Properties to see the information shown in Figure 13-11, which identifies the hinted streams and shows their respective data rates. Note that data rate is not one of the default columns; you have to right-click the window and choose it and other desired columns from the context menu.

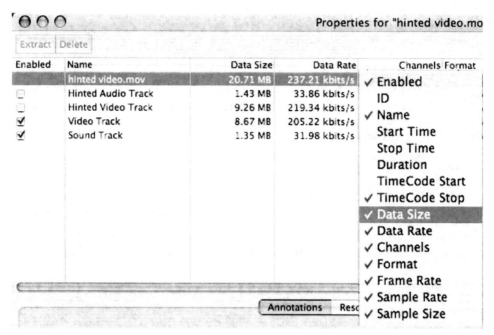

Figure 13-11. QuickTime Player's Movie Properties window.

Information about the hinted tracks proved particularly useful when producing with older versions of Telestream Episode, since the program allocates a disproportionately high data rate to the hinted video track—approximately 3 to 4 times higher than other encoders. For example, in Figure 13-11, you can see that the data rate for the hinted video track (219.34 kbits/s) is higher than that of the actual streamed video track (205.22 kbits/s). While this won't affect actual streaming bandwidth, since the hinted track never leaves the server, it does make the file look disproportionately large from a file-size standpoint. This tool was the only one that explained why the Telestream file was much larger than files produced by other encoding tools.

Other than this unique capability, however, QuickTime Player falls well short of other MOV analysis tools on the Windows and Mac platforms. Probably the best alternative is MediaInfo, which is similarly available on both platforms.

WMSnoop

Sliq Media's WMSnoop does only one thing—analyze Windows Media files—but it does that one thing very well. Throw in the free price tag and it's a must-have tool for anyone who creates, distributes or even just seriously watches Windows Media (**www.sliq.com/default.asp?view=wmsnoop**).

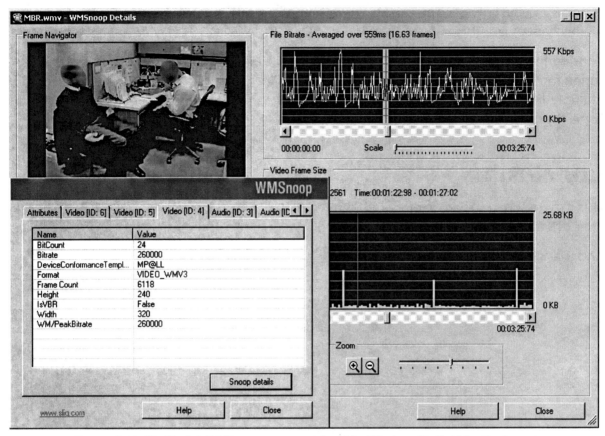

Figure 13-12. Sliq Media's WMSnoop, an essential tool for Windows Media Video Producers.

The program has at least three tabs for each WMV file: one for general attributes, one for video and one for audio. The Attributes tab contains a comprehensive list of file characteristics and, more importantly, which Windows Media Format SDK version was used to create the file. If you're wondering whether the file was encoded with SDK 10 or 11, this is one of the few ways to tell, and the only free one as far as I know.

Each Video tab contains the resolution and target bit rate for the stream, and the level and profile, but not the frame rate, though you can calculate this with information provided in other windows. The Audio tab contains all the expected data, plus it indicates whether the audio file was produced with VBR or CBR encoding. With multiple-bit-rate files, one tab details each audio and video stream, which you can't get elsewhere at any price.

If you click the Snoop Details button, you're rewarded with the graphic view on the lower right (Figure 13-12). You can gauge the variability of the data stream, identify I-frames (those tall frames on the lower right) or view the size of any frame in the file. I've used this to visualize the differences between VBR and CBR encoding, to compare third-party encoding tools and to see how the bit streams encoded by third-party vendors compares with those produced by

Windows Media Encoder. You can also determine if an encoding tool is inserting key frames at scene changes, another valuable quality-control metric.

Considering the price and utility, if you're a Windows Media producer who can't afford to spring for Semaphore, downloading this tool is a total no-brainer. If you're working with multiple-bit-rate files, WMSnoop provides specifics about the multiple streams that even Semaphore can't provide.

Conclusion

That's it. I'm sure there are some tools out there that I've missed, but hopefully you'll find the ones discussed of some benefit.

And that's it for this book. I hope you found it useful and perhaps even just a little bit fun.

Index

Symbols

A